WITHDRAWN

Communicating
in Geography
and the
Environmental
Sciences

OXFORD

UNIVERSITY PRESS

Oxford University Press is a department of the University of Oxford.

It furthers the University's objective of excellence in research,
scholarship, and education by publishing worldwide. Oxford is a registered
trademark of Oxford University Press in the UK and in certain other countries.

Published in Australia by
Oxford University Press
253 Normanby Road, South Melbourne, Victoria 3205, Australia

© Iain Hay 2012

The moral rights of the author have been asserted.

First published 1996
Second edition published 2002
Third edition published 2006
Fourth edition published 2012
Reprinted 2014, 2014(D)

National Library of Australia Cataloguing-in-Publication data

Author: Hay, Iain, 1960-
Title: Communicating in geography and the environmental sciences / Iain Hay.
Edition: 4th ed.

ISBN: 9780195576757 (pbk.)
Notes: Includes bibliographical references and index.

Subjects: Communication in geography.
Dewey Number: 910.72

Edited and proofread by Pete Cruttenden
Text design by Eggplant Communications
Typeset by diacriTech, India
Indexed by Russell Brooks
Cover image by Corbis/ Theo Allofs
Printed and bound in Australia by Ligare Book Printers, Pty Ltd

Communicating in Geography and the Environmental Sciences

Fourth Edition

IAIN HAY

OXFORD
UNIVERSITY PRESS
AUSTRALIA & NEW ZEALAND

CONTENTS

LIST OF FIGURES

LIST OF TABLES

LIST OF BOXES

ACKNOWLEDGMENTS

A large number of people have contributed to work that was eventually to become part of the pages which follow. I am very grateful to Cecile Cutler, Steve Fildes, Tom Jenkin and Noel Richards for their patient, thoughtful and constructive contributions to previous editions of this book. I am indebted to Stephen Fildes for his continuing hard work on the book's figures. And for this edition I really must commend Pete Cruttenden for his exemplary editing. I would like to thank Tania for reminding me to look at the garden and Della and Roley for making sure I spend time in it! I am also grateful to Taylor and Francis for allowing sections of articles published in the *Journal of Geography in Higher Education* since 1994 to be presented here in revised form. Full acknowledgments of those earlier works are included in this book's 'References and further reading' sections. Readers are also advised that with the exception of the chapter 'Writing a Media Release', material published in this book is reproduced to some extent in Hay, Bochner and Dungey's book *Making the Grade. A Guide to Successful Communication and Study* published by Oxford University Press. Finally, and again, to my parents—gifted teachers and learners—I owe special thanks.

INTRODUCTORY COMMENTS

Communicating in Geography and the Environmental Sciences, fourth edition, is about communicating effectively in academic settings. It discusses the character and practice of some of the most common forms of academic presentation skills used by students of geography and the environment. Chapters outline the 'whys' and 'hows' of essays, research and laboratory reports, reviews, media releases, summaries, annotated bibliographies, maps, figures, tables, posters, talks, examinations, referencing systems, and language and punctuation. Information on the ways in which these forms of presentation are commonly assessed is another important part of the book.

Knowledge, information and ideas remain the most highly valued currencies in universities. However, without the ability and means to communicate clearly and effectively, the value of one's thoughts and abstractions can be severely eroded. For that reason effective communication is a vital component of intellectual endeavour. One important ingredient of effective communication is an appreciation of the ways in which audiences make sense of the messages conveyed. Typically, audiences expect that certain conventions will be upheld or followed by people communicating to them through specific media. For instance, readers of an academic paper will usually expect some early introductory advice of the paper's purpose. People reviewing a scientific research report anticipate that information on supporting literature, research methods and results will be set out in a customary order and will offer specific sorts of information. Unfortunately, however, many students do not know the accepted cues, clues, ceremonies, conventions and characteristics associated with formal (academic) communication. In other words, some students do not know how to 'make the grade'. It is primarily because of this problem—and because of the lack of specific, relevant advice to address it—that this book was written.

In the pages which follow I have tried to demystify the conventions of communication associated with post-secondary education. By laying bare the criteria academic markers typically seek when evaluating specific forms of communication like essays, posters and talks, this book lets everyone know how grades are made. Just as importantly, the following pages set out the means by which grade-making criteria may be satisfied.

The book serves a number of other important purposes. It is intended to:

- *help improve teaching, learning and assessment.* Many academics now find themselves being asked to do more with less—to do more teaching, more research, more administration and more community service with less money, less time, less public recognition and less government support than ever before. One means of coping with this set of tensions is to teach more efficiently and effectively. This book is an attempt to contribute to those ends.
- *help increasingly diverse student populations fulfil educational objectives.* Recent moves to mass tertiary education in many Western countries have brought in students with far more diverse educational and cultural backgrounds than ever before. By revealing the characteristics of academic communication and assessment, this book represents one attempt to accommodate that shift. In serving this purpose, the book may also be of value to those offering and studying in distance education courses.
- *provide students with useful vocational skills.* In the context of emerging patterns of work and work organisation, the ability of university graduates to communicate ideas and information is becoming increasingly important; it also surfaces repeatedly in the reports of government and business think-tanks and academic authors. In some recognition of vocational considerations, this book makes an effort to contribute to the development of a range of communication skills.
- *codify and transfer teaching experience.* The book is one distillation of the experience and expertise possessed by university teachers who have, through both formal research and informal comment, shared their expertise with me on marking practice and communication conventions. With this book in hand, many new teachers and part-time teachers may find their teaching and assessment experiences simplified.

The first four chapters of the book focus on written communication. This includes material on essays, reports, annotated bibliographies, summaries, reviews and media releases. Emphasis in the next three chapters shifts to graphic communication, with specific attention given to posters, figures, tables and maps. These means of communication commonly are used—and referred to—by students of geography and the environmental sciences. The book then turns to spoken communication, and particularly to the 10–20 minute talk that is used with increasing regularly as an assessment task, in part because of its parallels with workplace experiences for many graduates of geography and the environmental sciences. The final two chapters take up subjects that intersect, one way or another, with most of the material set out earlier: exams and referencing. The first of these chapters deals with exams in their many forms: written, multiple-choice, oral, take-home and online. Depending on the type of exam, many of the skills discussed in previous chapters may need to be drawn upon. In the final chapter of the book, matters of referencing (and, to a lesser extent, language and punctuation) are taken up, these also being common to most of the forms of communication discussed earlier in the book.

Several of the chapters have been written around a common framework comprising four parts. First, there is an explanation of the specific type of communication being discussed. This takes the form of an answer to a question such as 'Why use a poster?' This section is followed by a broad, conceptual statement of the key matters assessors seek when marking the particular type of communication under consideration. For instance, lecturers marking book reviews typically seek clearly expressed description, analysis and evaluation of the text. An essay marker wishes to be told clearly what the author thinks and has learned about a specific topic. The third section consists of an outline of ways to achieve effective contact with one's audience. Where possible, this discussion is structured around an explicit statement of the sorts of criteria that assessors typically use when marking student work. A statement of those assessment criteria forms the fourth part of these chapters. Lecturers may find these lists useful as marking guides (for example, to ensure that a broad range of matters is considered during assessment, and to offer consistency in marking practice), elements of which might be compressed for practical purposes. They also may be of considerable benefit to students seeking either a checklist that might be used in critical self-review, or an indication of the sorts of things that are considered by assessors. Those chapters that are not structured following the framework set out above offer general advice on other critical matters, such as map making, citing sources and succeeding in exams.

... for students

If you are like many students in most universities, you have a mountain of textbooks to read, thousands of printed and photocopied words to digest, and collections of handwritten notes to absorb. On top of all that material directly related to your course, you now have a book on academic communication skills. Do not be discouraged. It is true that there is a lot to read and absorb in the pages that follow, but it is unlikely that you will have to use all of the material in this book in any single-semester university course. Instead, this is a book written to be used *throughout* your entire degree (and beyond). For example, in a first-year course, you might find the chapters on essay writing, graphics and exams most helpful. In third year, you might use the material on oral presentations for the first time, while further developing your skills gleaned earlier through the material on essays and graphics. You may even find sections of the book valuable if you go on to postgraduate study. So, think of the book as comprising sections that will be important over three or four years or even more, not just for one semester.

You might want to regard the book as something of a statement of achievement, too. Let me explain that rather cryptic comment. If you use this text in the next few years at university, you should be able to complete a degree with a sound appreciation of almost all of the material in the book. Look at the book as an indicator of proficiency, which says, 'This is what I will know', and not as something that says, 'This is what I have to know'.

The following pages discuss in detail the conventions of communicating effectively in an undergraduate academic setting. In part, the advice given is based on research into patterns of marking and comment by academic staff. The book was written to provide you with an insight into some of the expectations of people for whom you will be writing essays, giving talks and creating figures. Those expectations are reviewed fully in the chapters, but they are also summarised in the assessment schedules. An understanding of your audience's expectation *before* you undertake an assignment ought to help you prepare better work than might otherwise have been possible.

You should find it helpful to review the assessment schedules (and the relevant chapters) before you begin an assignment requiring some specific form of communication (for example, writing an essay or giving a talk). You will then be able to undertake the assignment with an understanding of the appropriate conventions of communication. When you have finished a draft of your work, try marking it yourself using the checklist as a guide. If you have a patient and thoughtful friend, ask them if they will do this for you, too. The checklist will help to ensure that you and your friend give consideration to the broad range of issues likely to be examined by any assessor, whether or not that assessor actually uses the schedules in their own marking practice. If something in the assessment schedule does not make sense to you, consult the material in the appropriate chapter for an explanation. In this way you may be able to illuminate and correct any shortcomings in your work before those problems are uncovered by your lecturer. The end result of this process ought to be better communication and better grades.

One thing needs to be stressed from the outset. The following pages are intended to offer advice only and are not prescriptions for 'perfect' assignments. The guidelines are intended to assist you in preparing assignments, types of which you might be undertaking for the first time (for example, an academic poster or a formal talk). As well as heeding the content of this book, you should read journal articles and other people's essays critically, pay attention to the ways in which effective and poor speakers present themselves and their material, and critically assess maps and graphs. See what works and what does not. Learn from your observations and attempt to develop your own distinctive approach to communicating. The style you develop may be very effective and yet transgress some of the guidelines set out in the following pages. This should not be a matter of concern: your individuality and imagination are to be celebrated and encouraged, not condemned and excised.

And if the content of this book still leaves you with questions about communicating in geography and the environmental sciences, there are three additional places to seek help. First, consult some of the many guides to scholarly communication in your library or online. Second, if the institution in which you are studying has a student support centre or study skills centre, speak to the staff there. Third, ask your tutor or lecturer for help. Although writing centre and academic staff will probably be less than willing to act as your personal editor, they will probably be very happy to offer general guidance and to answer any specific questions you might have.

... for lecturers

Despite the revisions, this fourth edition is still not a book of magic. I do not wish to make wild claims about the benefits that might flow from student and teacher use of the pages that follow. However, on the basis of experiences with the earlier editions, I do have reason to believe that if material from this book is referred to and incorporated into teaching *and* assessment practice within a discipline or across a degree, there is likely to be an improvement in student communication skills. The following simple and effective strategies for using this book have yielded positive results.

- Before the class undertakes an exercise, make available to students relevant copies of assessment schedules associated with most chapters. Alternatively, ask students as individuals or in groups to prepare their own criteria for establishing whether a specific piece of work is successful. Discuss these criteria within the class and compare them with the assessment schedules set out in this book. Perhaps produce a composite set of criteria that acknowledges the intentions and ambitions of both yourself and your students. In other words, use the assessment schedules provided here as a means of encouraging your students to think about their audience's expectations. This can also help you to simplify (where you consider it appropriate) the sometimes extended assessment schedules set out in this book.

- Encourage students, by whatever means you consider appropriate, to read assessment sheets and explanatory notes in the appropriate chapter of the book critically before they begin an assigned task.
- Apply strategies that encourage students to use marking schedules to assess their own work before submitting it for peer review or final assessment. Not only does this offer an opportunity for students to engage in critical self-reflection, but if submitted with work for instructor assessment, self-assessment sheets also can be used as a 'diagnostic' tool, highlighting differences between students' perceptions of their own work and those of the assessor.
- Use peer review methods such as writing groups and student assessment of oral presentations as means of encouraging students to think critically about communication and the positions and expectations of author and audience. The assessment sheets associated with several of the chapters in this book are useful preliminary frameworks for peer review.
- Use the assessment forms, and use them repeatedly, as the foundation for your own assessment practice.

These strategies, used singly and in combination, require little extra teaching effort but offer the potential for improvements in written, oral and graphic communication skills. Why not give them a try?

ONE

Writing an essay

True ease in writing comes from art, not chance,
As those move easiest who have learned to dance.

Alexander Pope

Key topics

- Why write?
- What is an essay?
- How do I write a good essay?
- What are your essay markers looking for?

This chapter briefly **argues** the case for writing, before going on to **discuss** how to write a good **essay**. Most of the material in the chapter is devoted to a **review** of those matters your essay markers might be looking for when they are **assessing** your work. Much of the information and advice in the following pages is structured around the essay assessment schedule included at the end of the chapter.

Why write?

You might think that essays and other forms of written work demanded by your lecturer are some sort of miserable torture inflicted on you as a part of an ancient academic initiation ritual. To tell the truth, however, there are some very good reasons to develop expertise in writing.

- It is an *academic and professional responsibility* to write. As an academic or practising geographer (perhaps labelled economist, planner or demographer) or environmental manager you should make the results of your work known to the public, to government, to sponsoring agencies and the like. There is not only a moral obligation to make public the results of scientific inquiry, but you will also probably be required to write—and to write well—as part of any occupation you take up.
- Writing is one of the most *powerful means we have of communicating*. It is also the most common means by which formal transmission of ideas and arguments is achieved.
- Among the most important reasons for writing is the fact that writing is a *generative, thought-provoking process*:

  I write because I don't know what I think until I read what I have to say.

  Flannery O'Connor

  You write—and find you have something to say.

  Wright Morris

  But I really write to find out about something and what is known about something ... I write books to find out about things.

  Dame Rebecca West

As these quotes suggest, writing promotes original thought. It also reveals how much you have understood about a particular topic. (For a discussion of ideas on this point, see Game and Metcalfe's (2003) chapter on writing.)

- Writing is also a means by which you can *initiate feedback on your own ideas*. Through the circulation of your writing in forms such as professional reports, essays, letters to the editor and **journal** or magazine articles, you may spark replies that contribute to your own knowledge as well as to that of others. Writing (as well as other forms of communication) is critical to the development and reshaping of knowledge.
- By forcing you to marshal your thoughts and present them coherently to other people, writing is also a *central part of the learning process*.
- Writing is a means of *conveying and creating the ideas of new worlds*. Writing is part of the process by which we give meaning to, and make sense of, the

world(s) in which we live. People who write well, including some journalists, can influence how we think about important matters. Therefore, learning to write well offers some capacity to control our destiny.

- *Writing can be fun.* Think of writing as an art form or as storytelling. Use your imagination. Paint the world you want with words.

What is an essay?

In educational contexts, an essay is a concise, organised, written discussion of your considered ideas on a specific topic. It is commonly based on a synthesis of evidence and ideas drawn from previously published sources (for example, journal articles, books and government reports) and supported by examples obtained from those sources. Essays are often used by lecturers as assessment tools to develop and judge your mastery or comprehension of material, as well as your ability to communicate ideas clearly in written form. Different types of essay exist (David & Liss 2006; Henderson & Moran 2010), including cause and effect essays, argumentative essays and reaction essays. Less important here than a categorisation and description of essay types is the point that an essay should tackle the assigned topic in the assignment's own terms. So, for example, if you are asked to 'argue the case for or against increased levels of taxation spending on public transport in Sydney' (an argumentative topic), you should not tackle it as a review essay in which you appraise the relative merits of past arguments for or against such increased expenditures.

How do I write a good essay?

During your degree program, you might write twenty to fifty essays, totalling about 100,000 words—more words than there are in this book! You might as well spend a little time now to ensure that the 1000 to 2000 hours you spend writing those essays are as productive and rewarding as possible.

- *Read widely.* Reading the work of others—especially geographers and environmental scientists—is essential. Not only will it increase your knowledge of your subject, but it will also give you a feeling of how experts write about it.
- *Devote sufficient time to research and writing.* There is no formula for calculating the amount of time you will need to devote to writing essays of any particular length or 'mark value'. Some writers do their best work under great pressure of time; others work more slowly and may require

several weeks to write a short essay. However, regardless of your writing style, doing the research for a good essay does take time. So, do yourself a favour and devote plenty of time to finding and reading books and journal articles, and to consulting other sources germane to your essay.

- *Practise your writing.* With most art forms, just as with sport, practice improves your ability to perform. Practice allows you to apply the 'conventions' of effective writing. It also offers the opportunity to seek feedback on the quality of your writing.
- *Plan your writing.* Plan your essay's structure before you begin writing, and devise a work schedule that allows sufficient time for writing the essay.
- *Write freely at first, but work within the structure or plan you have devised for your essay.* Suspend **editing** until you have a significant first draft of whatever part of your essay you are working on (Greetham 2008). It is much easier to edit and polish a draft than it is to try to prepare a final version on the first attempt.
- *Edit and revise.* Perhaps you have an image of a good writer sitting alone at a keyboard typing an error-free, comprehensible and publishable first draft of a manuscript. Sadly, that image is unfounded: almost every writer produces countless drafts. Rewrite and rewrite again, being sure to give yourself the opportunity to review your essay after some time (days, not minutes) away from it.
- *Seek and apply feedback.* Writing is an individual yet *social* process. Many successful writers seek **comment** from peers and other reviewers to find out how the words they have written are interpreted by others. This is known as **peer review**. Listen to your readers' comments carefully, but remember that in the end that it is *your* essay.
- *Read your 'final' essay aloud.* No matter how many times you and your friends read over your work in silence, it will almost always be improved by correcting it after hearing it read aloud. Spoken aloud, you will hear repetitions in your work that you did not know were there. Unclear ideas and **sentences** will reveal themselves, and pompous, grandiloquent language will become evident.

What are your essay markers looking for?

Many students waste a great deal of time trying to work out what specific 'angle', approach or 'line' their lecturer is looking for in an assignment, under the misapprehension that careful adoption of that approach will lead to a high grade. In essays, your lecturers are not looking for 'correct answers'. There is no particular 'line' for you to follow. Lecturers are concerned with *your* views and how well you make *your* case. Whether they agree or disagree with your judgment is not essential to your mark. Disagreement does not lead to bad marks; bad essays do (Lovell & Moore 1992, p. 4).

As Box 1.1 makes clear, the answer to the question, 'What are your markers looking for?' is really quite simple.

- to be told clearly what you *think*
- and what you have *learned*
- about a *specific* topic.

> **Box 1.1**
> **Essay assessors want …**

The following guidelines and advice, which match the criteria outlined in the assessment schedule at the end of this chapter, ought to help you satisfy the broad objective noted in the shaded box above. It is worth thinking seriously about these guidelines.

Quality of argument

Ensure that the essay fully addresses the question

If any one issue in particular can be identified as critical to a good essay, it is this one. Failure to address the question assigned or chosen is often a straightforward indicator of a lack of understanding of the course material. It may also indicate carelessness in reading the question or a lack of interest and diligence.

- *Look closely at the wording of your essay topic.* For example, what does '**describe**' mean? How about '**analyse**' or '**compare** and **contrast**'? What do other key words in the assigned topic actually mean? (See the glossary for a discussion of terms used commonly in essay assignments.) The difference in meaning can be critical to the way you approach your essay. For example, an essay in which you are asked to 'critically discuss John Howard's role in post-2000 Australian race relations' requires more than a description of what he did. It asks you to **evaluate** the significance of his actions. Similarly, an essay that asks you to 'discuss the implications of tourism for the cultural integrity of the indigenous inhabitants of Tonga' requires that you give central attention to the issue of cultural integrity and not that you discuss the history of tourism in Tonga. You should be aware that most of your university essays will require you to think about material and weave it into an **argument**. Very few will ask you to simply recount all the facts you have discovered about some phenomenon or issue.

- *Discuss the topic with other people in your class.* See how your friends have **interpreted** it. Listen critically to the views of others, but be prepared to stand up for your own and change them only if you are convinced you are wrong.
- *Wherever necessary, clarify the meaning of an assigned topic with your lecturer.* Do this *after* you have given the topic full thought, discussed it with friends and established your own interpretation, but before you begin writing your paper. Lecturers do get frustrated with students who seek guidance without having first sought to resolve the challenge for themselves.
- *When you have finished writing, check that you have covered all the material required by the nature of the topic.* The paragraph listing and rearranging technique outlined in the next section of this chapter may be helpful.

Ensure that your essay is logically developed

Nothing is more frustrating than to be lost in someone else's intellectual muddle. A paper that fails to define its purpose, that drifts from one topic to the next, that 'does not seem to go anywhere,' is certain to frustrate the reader. If that reader happens to be the person marking your work, he or she is likely to strike back with notations scrawled in the margin criticising the paper as 'poorly organised,' 'incoherent,' 'lacking clear focus,' 'discursive,' 'muddled,' or the like. Most lecturers have developed a formidable arsenal of terms that express their frustration at having to wade through papers that … are poorly conceived or disorganised.

Friedman & Steinberg (1989, p. 53)

In assessing an essay, markers will usually look for a coherent framework of thought underpinning the work. They are trying to uncover the conceptual skeleton upon which you have hung your ideas and to see if it is orderly and logical. Throughout the essay, readers need to be reminded of the connections between your discussion and the framework. Make clear the relationship between the point being made and any argument you are advancing.

For most topics, an essay framework can be formed before writing begins (see the **essay plan** and sketch diagram discussions in Box 1.2). In other situations, the essay may take shape as it is being written (see **freewriting** in Box 1.2).

Essay plan

Try to set out an essay plan before you begin writing. That is, work out a series of broad headings that will form the framework upon which your essay will be constructed. Then, add increasingly detailed material under those headings until your essay is written. What will be the main ideas you might cover? What examples, data and quotes might be useful? What **conclusions** might you reach? As you proceed, you may find it necessary to make changes to the overall structure of the essay.

Sketch diagrams

A sketch diagram of the subject matter is also a good means of working out the structure of your essay. Write down key words associated with the material you will discuss, and draw out a sketch of the ways in which those points are connected to one another. Rearrange the diagram until you have formulated an outline. This can then be used in much the same way as an essay plan.

Freewriting

If you encounter 'writer's block', or are writing on a topic that does not lend itself to use of an essay plan, brainstorm and without hesitation write anything related to the topic until you have some paragraphs on the screen or page in front of you. Then remove the rubbish and organise the material into some coherent package, taking care to eliminate repetition. To be effective, this writing style requires good background knowledge of the material to be discussed. Freewriting can be a good way of making connections between elements of the material you have read about. It is not an easy option for people who do not have a clue about their essay topic.

Box 1.2
Working out a structure for your essay

Plan your essay.

When you have written the first draft of your essay, check the structure. You can do this quite easily:

- Go through the document giving each **paragraph** or section a heading that **summarises** that section.
- Write out the headings on a separate sheet of paper or on cards, or use the 'View Outline' option in your **word processing** software to list the headings used throughout the document. Read through the headings. Are they in a logical order? Do they address the assigned topic in a coherent fashion?
- If necessary, rearrange the headings until they do make sense and add new headings that might be necessary to fully cover the topic. If additional headings are required, you will also have to write some new sections

of your paper. Of course, you may also find that you can remove some sections.

- Make your amendments and then go through this process of assigning and arranging headings again until you are satisfied that the essay follows a logical progression.
- Finally, rearrange the written material according to the new sequence of headings.

Make sure your essay structure is clear and coherent.

You might find that the summary of headings you have prepared supplies a framework upon which you can develop an informative **introduction**. A reader provided with a sense of direction early in the essay should find your work easy to follow.

Ensure the writing is well structured through introduction, body and conclusion

Imagine your essay has an hourglass structure.

In almost all cases, good academic writing will have an introduction, a discussion and a conclusion. It is helpful to visualise this structure in the form of an hourglass. The introduction provides a broad outline, setting the topic in its context. The central discussion tapers in to cover the detail of the specific issues you are exploring. The conclusion sets your findings back into the context from which the subject is derived and may point to directions for future inquiry.

The guidelines in Box 1.3 are not suggested as a recipe for essay writing, but should be of some assistance in constructing a good paper.

Box 1.3
Guide to the contents of an essay's introduction, body and conclusion

In the introduction

- **State** your aims or purpose clearly. What problem or issue are you discussing? Do not simply repeat or rephrase the essay question: that is a sure-fire way of putting any reader off your work.
- Make your **conceptual framework** clear. This makes it easier for readers to understand the rest of your essay.
- Set your study in context. What is the significance of the topic?
- Outline the scope of your discussion (that is, give the reader some idea of the spatial, temporal and intellectual boundaries of your presentation). What case will you argue?
- Give your reader some idea of the plan of your discussion: a sketch map of the intellectual journey they are about to undertake. Leave them in no doubt that your essay has a clear and logical structure (Burdess 1998).
- Be brief. In most essays an introduction that is about 10 per cent of the total essay length is adequate.

Good essays get off to a good start.

- Capture your reader's attention from the outset. Is there some unexpected or surprising angle to the essay? Alternatively, you might capture attention with relevant and interesting quotes, amazing facts and anecdotes. Make your introduction clear and lively, because first impressions are very important.

The best introductions are those that get to the point quickly and that capture the reader's attention (May 2010). Like a good travel guide, an effective introduction allows readers to **distinguish** and understand the main points of your essay as they read past them (Greetham 2008).

In the discussion

- Make your case. 'Who dunnit?'
- Provide the reader with reasons and **evidence** to support your views. Imagine your lecturer is sitting on your shoulder (an unpleasant thought!) saying '**prove** that' or 'I don't believe you'. Disarm his or her scepticism.
- Present your material logically, precisely and in an orderly fashion.
- Accompany your key points with carefully chosen, colourful and correct examples and analogies.

Convince your reader with logic, example and careful structure.

In the conclusion

- State clearly your resolution of the problem or question set out in the introduction. The conclusion ought to be the best possible answer to your essay question on the basis of the evidence that you have discussed in the main section of the paper (Friedman & Steinberg 1989, p. 57). It must match the strengths and balance of material you have presented throughout the essay (Greetham 2008).
- Avoid clichéd, phoney or mawkish conclusions (Northey et al. 2009, p. 85). For example, 'The tremendous amount of soil erosion in the valley dramatically highlights the awful plight of the poor farmers who for generations to come will suffer dreadfully from the loss of the very basis for their livelihood.'
- Do not introduce new material or ideas that have not been presented earlier in the essay (Greetham 2008).
- If it is appropriate, and if you can, leave your reader with something interesting to think about (Najar & Riley 2004, p. 56), such as wider implications or your informed perspective on future trends. For instance, to conclude an essay that has reviewed recent trends in geographical thought, you might speculate briefly on the way theory may take shape over the next few years.
- Tie the conclusion neatly together with the introduction. When you have finished writing your essay, read just the introduction and the conclusion. Do they make sense together?

Make sure the conclusion matches up with the introduction.

- Finally, ask yourself: have I answered the question? This may be evident only if you have the opportunity to review your essay some days after you have finished writing it. For this reason, if for no other, it is a good idea to plan to finish the penultimate draft of your essay some days or weeks before the due date. This will give you the opportunity to more dispassionately review your own work.

Headings show structure.

Some writers like to use *headings* throughout their essay. You are not compelled to use headings in essays; indeed, some lecturers actively discourage their use (check with your lecturer). However, for readers of an essay, headings may do several things (Goldbort 2006, p. 142; Snooks & Co. 2002, p. 44; Windschuttle & Elliott 1999). Headings can help to: map the essay's structure; **show** readers where to find specific information within the essay; group information into clearly defined sections; and **indicate** to the reader what is to follow. These points provide a clue about the number and nature of headings that might be included in an essay. Provide enough headings to offer a person quickly scanning the essay a sense of the work's structure or intellectual 'trajectory'. To check this, write out the headings you propose to use. Is the list logical or confusing, sparse or detailed? Referring back to the essay itself, revise your list of headings until it provides a clear, succinct overview of your work.

If you are writing a particularly long essay, you may need a hierarchy of headings. Three levels should be sufficient for most purposes, although up to five are **illustrated** in Box 1.4. Too many headings can add confusion rather than clarity.

Box 1.4 Heading styles

BOLD CAPITALS

MEDIUM CAPITALS

MEDIUM ITALIC CAPITALS

CAPITALS AND SMALL CAPITALS

or

BOLD CAPITALS

Bold lower case

Bold italic lower case

Italic lower case

Italic lower case. With text running on ...

or

BOLD CAPITALS

Bold Upper and Lower Case

Bold italic lower case

Italic lower case

Take care to ensure that section headings are consistent throughout the essay.

Ensure the material is relevant to the topic

The material you present in your essay should be clearly and explicitly linked to the topic being discussed. To help clarify whether or not material is relevant, try the following exercise. When you have finished writing a draft of your essay, read each paragraph asking yourself two questions:

Don't waffle!

1 Does *all* of the information in this paragraph help answer the question?
2 *How* does this information help answer the question?

On the basis of your answers, edit your work further. This should help you to eliminate the dross.

Ensure the topic is dealt with in depth

Have you simply slapped on a quick coat of paint or does your essay reflect preparation, undercoat and good final coats? Have you explored all of the issues emerging from the topic? This does not mean that you should employ the 'shotgun' technique of essay writing. This poor essay-writing style sees the author indiscriminately put as much information as they can collect on a subject onto the pages of their essay.

Instead, be diligent and thoughtful in going about your research, taking care to check your institution's library, statistical holdings, other libraries and information sources. Take notes and make photocopies. Read. Read. Read. There is not really any simple way of working out whether you have dealt with a topic in sufficient depth. Perhaps all that can be said is that broad reading and discussions with your lecturer will provide some indications.

Quality of evidence

Ensure your essay is well supported by evidence and examples

You need relevant examples, statistics and **quotations** from books, articles and interviews, as well as other forms of evidence to support your case and substantiate your claims (for a discussion of evidence, see Chapter 7 of Henderson & Moran 2010). In addition, most readers seek examples that will bring to life or emphasise the importance of the points you are trying to make. You can draw information from good research (for example, by reading widely or by conducting interviews with appropriate people). Careful use of examples is also an effective indicator of diligence in research and of the ability to link **concepts** or theory with 'reality'.

When looking for evidence to include in your essay, do not confine your search to material available through the **World Wide Web**. Your essay

Use sources other than the World Wide Web.

marker will most likely be very disapproving of this. Although the web is a remarkable and increasingly valuable source of information (see, for example, useful resources like Google Scholar and JSTOR that provide free access to scholarly works), using it solely and fundamentally sometimes indicates lazy scholarship and neglects vast amounts of high-quality material available from other sources—notably academic journals, reputable newspapers and magazines, and books.

Make sure all of your sources are credible.

Be critical of *all* your sources and particularly those on the internet. Whose evidence and argument do you accept? Why? How do you know that your source is in any way credible or authoritative? Whose interests does it represent? By way of a cautionary example, **consider** anwr.org, which is a website that promotes oil drilling in the Arctic National Wildlife Refuge. It takes some investigation to find that the site is supported by Arctic Power, a coalition of Alaskan industry groups pressing to open parts of the refuge to oil and gas development. Arctic Power is underwritten by the state of Alaska with funding from the oil industry.

Table 1.1 provides some suggestions for scrutinising the credibility of **web pages**. Simply because something is published, or is on the web, does not mean it is true. Indeed, as a 1993 Peter Steiner cartoon published in *The New Yorker* magazine observed: 'On the internet, nobody knows you're a dog.'

Used judiciously, personal experience can be good evidence for an essay.

Personal experience and observations may be incorporated as evidence in written work. For example, if you have spent several years as a police officer you may be able to offer penetrating insights in an assignment on the geographies of crime and justice. Women and men who have spent time caring for children in new suburban areas may have valuable comments to make on the issue of social service provision to such areas. It is certainly valid to refer to your own experiences, but be sure to indicate in the text that it is to those that you are referring, and provide the reader with some indication of the nature and extent of your experience; for example, 'In my nineteen years as a police officer in central Melbourne … '. Where possible, support your personal observations with other sources that readers may be able to consult.

Avoid making unsupported **generalisations**; for example, 'Crime is increasing daily' or 'Air pollution is the major cause of respiratory illness'. Unsupported generalisations are indicators of laziness or sloppy scholarship and will usually draw criticism from essay markers. Provide support for claims through use of empirical evidence or by citing recognised sources; for example, 'The Australian Bureau of Statistics has recently released a report entitled *Crime in Australia* that **demonstrates** that crime rates in Australian urban areas have risen since comparable figures were last collected', or 'A new paper in the medical journal *Lancet* suggests that air pollution can be linked to 1054 deaths in Mexico City since 2010'.

Table 1.1 Evaluating web pages

Question	How do you answer it?	What does it imply?
Did you find the page or site through a sponsored link?	Some **search engines** will direct you to sponsored links first. These are usually identified as such.	Information distributed with commercial intent is not always 'balanced'.
From what domain does the page come?	Look at the **URL (Uniform Resource Locator)**. Is the domain, for example: • commercial (.com) • educational (.edu) • government (.gov) • non-profit (.org) • miscellaneous (.net)? What country is it from? For example, .au for Australia, .nz for New Zealand or .ca for Canada (note that USA-based sites typically have no country identifier). A web search for country **domain names** will yield full lists. Consider whether the domain and country seem appropriate for the site	Consider the appropriateness of the domain to the material. Is this kind of agency a fitting one for the material being presented? Is the material from the right place?
Who or what agency wrote the page and why? Is the agency reputable?	Good places to start are the banner at the top of the page or any statement of copyright, which is typically located at the bottom of the page. Alternatively, look for information under links associated with the page and characteristically entitled 'About us', 'Who we are', or 'Background'. Authoring agency details are sometimes located in the URL between the http:// statement and the first / (forward slash), or immediately after a www statement (e.g. www.abs refers to the Australian Bureau of Statistics and www.maf refers to the NZ Ministry of Agriculture and Fisheries). Try truncating the URL to find out about the authoring agency. That is, starting from the end of the URL, progressively delete each phrase ending with a / (forward slash), pressing enter after each deletion. This may generate new web pages that provide insights into the origins of the page you planned to use.	Web pages are written with intent or purpose—and not always with the best of intentions!

Table 1.1 (*cont.*)

Question	How do you answer it?	What does it imply?
Who or what agency wrote the page and why? Is the agency reputable?	It may also be useful to query the authoring institution's name through a search engine. This may reveal other information that points to funding sources and underlying agendas	
Is this someone's personal web page or part of a **blog (weblog)***?*	Look for a personal name in the URL. This is typically shown in the URL after a ~(tilde), % (percentage symbol) or /people/ or /users/ statement (e.g. /~jtrout/). Blogs are often identified as such through their title (e.g. Wired Campus Blog), through their URL (which may contain the word 'blog') or in the web page text that introduces the blog.	The fact that information is presented on a personal page or 'blog' is not necessarily a bad thing. Indeed, some are of great value being written (e.g. by people with media connections but without the oversight of an editor). However, you will need to find out whether the author is an expert or credible source. This information may be set out on the site you are looking through. If not, use a good search engine (e.g. Alexa, Bing, Google or Yahoo!) to query the author's name. If you cannot get any insights to the author (e.g. their credentials or professional role), think seriously about whether you should use information from the page.
When was the page created or last updated?	Look at the bottom of the web page. This is usually where a 'created on' or 'last updated on' statement is located.	Old pages may contain outdated information. In almost every case, undated statistical or factual information should not be used.
Is the content and layout of high quality?	Check to see if the page looks well produced and that the text is free of typographical errors and spelling mistakes (translated foreign sites may be an exception). Where possible, confirm the plausibility and accuracy of data or other information presented by comparing it with other good sources.	Scruffy, poorly set-out web pages do not necessarily contain inaccurate information, but they should cause you to question the meticulousness of the author in their information gathering and presentation.

Adapted from: Barker (2007) and Beck (2009).
Barker, in particular, offers very useful and detailed advice on assessing web pages.

Ensure accurate presentation of evidence and examples

When you use examples, make sure they:

- are relevant
- are as up-to-date as possible
- are drawn from trustworthy sources such as official statistics, *Hansard*, well-regarded journals, high-quality **websites** or experts in the field (all fully identified in your text with an appropriate referencing system)
- include no errors of fact.

Keep a tight rein on your examples. Use only those details you need to make your case.

Use of supplementary material

Make effective use of figures and tables

You can use illustrations to make points more clearly, effectively or succinctly than you can make them in words. People usually remember the information in illustrations more easily than that in text. **Histograms**, **pie charts**, tables, schematic diagrams and photos can supplement text, but should not duplicate it (Mullins 1983). Illustrations can be drawn from your reading—with acknowledgement, of course—or you can create your own where appropriate (see Chapters 6 and 7 for details).

When evaluating your use of illustrative material, such as figures, tables and maps, markers will check to see that you have made reference to the illustration in your discussion and that the illustration makes the point intended. You can emphasise especially important elements of an illustration in your text, but avoid setting out a detailed description of all its contents. Assessors also look to see whether you might have added additional illustrative material to support the points you are making or to better organise the information you have presented.

A picture is worth a thousand words.

You should incorporate tables and graphics into your essay with care. Make sure they are relevant and locate them as closely as possible to that part of the text in which they are discussed. Be sure to discuss each illustration somewhere in your text. Unless you have a particularly good reason for doing so, do not put figures in an **appendix** at the end of the essay. Most readers find this very frustrating.

Ensure illustrations are presented correctly

Several types of illustrative material are commonly included in written work. These, and the accepted names for each, are listed in Table 1.2.

Tables, figures and plates can contribute substantially to the message being communicated in a piece of work. However, you must take care with their presentation.

Consult Chapter 6 in this book for more information on the presentation of figures and tables, and Chapter 7 for details on making a map.

**Box 1.5
Using
illustrative
material**

Illustrative material should be:

- customised to your work. Do *not* submit an essay laden with marginally relevant photocopied tables and figures lifted directly from texts and journals or pasted in from the web. Where appropriate, redraw, rewrite or modify the material to suit your aims.
- legible and large enough to show *all* key details clearly. Do not include blurry or unreadable images.
- comprehensible. Is the illustration easily understood and self-contained? Maps and other diagrams should have a complete and comprehensive *key* or *legend* that allows readers to comprehend or decode the material shown. *Labelling* should also be neat, legible and relevant to the message being conveyed by the illustration.
- correctly identified with sequential Arabic numerals beginning with 1. For example, you should not photocopy Table 17.1 from a textbook and insert it in your essay as Table 17.1. Give it a number customised to your work (begin with Table 1 or Figure 1 etc.) and remove all trace of the photocopied numbering. When you place illustrative materials in your work, ensure that tables, figures etc. are put in correct numerical order. For example, Figure 3 should precede Figure 4.
- clearly and comprehensively titled. The title must fully specify the *subject* of the illustration, its *location* and the *time period* to which it refers. For example, 'Vietnamese-Born Population as Percentage of Total Population, Adelaide Statistical Division, 2012' is a good title, whereas 'Vietnamese Population' is not.
- properly cited. The *source* from which you derived the illustration should be specified. Failing to correctly identify the source is one of the most common problems associated with student use of illustrations in written work. The source should be acknowledged with an appropriate reference (see Chapter 10 for more details).

Table 1.2 Types of illustrative material

Type of material	Label
Graphs, diagrams and maps	Figure
Tables and word charts	Table
Photographs	Plate or Photo

Written expression and presentation

Your writing should be fluent and succinct

Write simply—short words and sentences are best—and limit your use of **jargon**. Unless you deliberately wish to obfuscate, there is little room for grandiloquence in effective written communication. (I trust you see the irony in that sentence.) In your essay you should be conveying knowledge and information, not showing how many big words you know (Booth 1985, p. 13). One straightforward means of checking that communication is clear is to ask yourself whether your writing would be understood by someone whose first language is not English. If you know someone in that position who is willing to proofread your essay, give them a copy to look over.

Another very effective means of checking the fluency of your writing involves putting a draft of your work away for several days and then reading it afresh. Odd constructions and poor expression that were not evident before will leap out to greet you. This exercise is often even more revealing if you read the essay aloud. Yet another alternative is to make some mutual editing arrangements with friends (taking care, of course, not to violate any of your university's rules on authorship). Make a deal that you will 'correct' their papers if they will 'correct' yours. Booth (1985, p. 6) notes that for over 2000 years it has been known that we see other people's mistakes more easily than we see our own.

Reading your essay aloud can reveal clumsy expression, poor punctuation and repetition.

Learning fluency in writing may seem impossible. There are, however, two fundamental secrets to success:

1 Take care to keep sentences short. Short sentences are easy to read. Short sentences convey ideas in a no-nonsense style.
2 Effective paragraphing is important. Aside from signifying a subdivision within your text that that is greater than the division between sentences, paragraphing offers your reader visual clues to separations of subject matter that can make reading easier. Although there are exceptions (see Clanchy and Ballard 1997 for a discussion), paragraphs typically comprise three parts (Barrett 1982, p. 118), as illustrated in Box 1.6.

**Box 1.6
Parts of a
paragraph**

- The **topic sentence** states the main idea in what is usually the most general sentence of the paragraph; for example, 'The depletion of Brazil's tropical rain forests is proceeding apace'.
- **Supporting sentences** offer answers to 'why' and 'how' questions, and use examples to support the topic or to prove the point; for example, 'There is little government action to end land clearance in fragile environments and private incentives to clear the land remain attractive'.
- The **clincher** lets the reader know the paragraph is over. It may summarise the paragraph, echo the topic sentence or ask a question; for example, 'There seems to be little hope for the sustainability of forests of the Amazon region'.

Most paragraphs are unified by a *single* purpose or a single theme (Moxley 1992, p. 74). That is, a paragraph is a cohesive, self-contained expression of one idea. If your paragraph conveys a number of separate ideas, rethink its construction, and think about dividing long paragraphs into shorter ones.

Successive paragraphs should **relate** to one another as well as to the overall thrust of the text. Get into the practice of using transitional sentences at the beginning or end of paragraphs to make clear the connections between the material discussed in each. Following are examples of common transitional phrases that serve that purpose:

- Another problem associated with ...
- On the other hand ...
- Furthermore ...
- A similar explanation ...
- From a different perspective ...
- Elsewhere ...
- Other common ...
- As a result ...
- A significant consequence is ...
- A number of issues can be identified ...

These are devices that allow you to lead your reader from one part of your essay to another.

Make every word count. Waffle is easily detected and it makes assessors suspect that you have little of substance to say. Prune unnecessary words and phrases from your work. Remember that the key objective in an essay is to answer the question or to convey a body of information—not to write some minimum number of words!

Use grammatical sentences

One simple way of detecting difficulties with grammar is to have that trusty friend read your essay out loud to you. If that person has difficulty and stumbles over sentence constructions, it is likely that the grammar is in need of repair. Another simple means of avoiding problems is to keep sentences short and simple. Not only are long, convoluted sentences often difficult to understand, but they are also grammatical minefields.

Use correct punctuation

Check the material in Chapter 10 for a review of common punctuation problems. Take particular care with the use of apostrophes! If there is anything you do not understand, check an **online** punctuation guide, ask your lecturer or speak to the good folk at your university's study skills centre.

Use correct spelling throughout

Incorrect spelling brings even the best of work into question. Spelling errors emerge repeatedly as a problem in university-level essays. Given that many papers are written with the help of a word-processing package, there is little excuse for incorrect spelling. Be sure to use the spell checker before you submit your essay for assessment, but remember that distinctions between words such as there/their, too/two/to and bough/bow will not show up. Also check that your spell checker is using the appropriate form of English (for example, Australian English rather than US English or Philippine English). And it can be very useful to have a good dictionary available. Spell checkers and word-processing thesauri will not always have the answers you seek.

Spelling errors and sloppy presentation reflect poorly on your work.

Make sure your work is legible and well presented

> Poor presentation can prejudice your case by leading the reader to assume sloppiness of thought.
>
> Bate & Sharpe (1990, p. 38)

Essays that are difficult to read because of poor presentation can infuriate assessors because it can be difficult to maintain a sense of your case, argument or evidence if reading is disrupted by bothersome matters of appearance. It is now generally expected that essays will be written on computers and this has done much to eliminate legibility challenges presented by many handwritten essays. However, a few new presentation matters have emerged. To help overcome these, many universities now have uniform requirements surrounding the appearance of essays. In the case that yours does not, use a 12-point, Times New Roman font, with 1.5 or double spacing and wide margins all around (3.5–4 cm). This makes your work easier to read and allows some space for your marker's comments.

Nicely presented work suggests pride of authorship. You are likely to find that presentation does make a difference—to your own view of your work as well as to the view of the assessor.

Your assignment should be a reasonable length

A key to good communication is being able to convey a message with economy (consider the communicative power of some short poems, such as *haiku*). Take care not to write more words than have been asked for. Most people marking essays do not want to read any more words than they have to. Avoid masking a scarcity of ideas with verbose expression, padding out your essay with filler material or 'waffle'. If you find you do not have enough to say in an essay, perhaps you need to do some more research—not more writing.

Sources and referencing

Keep a full record of all your sources. It will save you a lot of time and work when you prepare your final reference list.

Two simple undertakings can save you a great deal of anguish in essay writing. First, keep a full record of the **bibliographic details** of all the **references** you use. This includes, but is not limited to, information such as: who is the author; when was the work published; where; and by whom? Details of what constitutes a full record of bibliographic details is set out in Chapter 10. It can be helpful to photocopy the title page of articles (if they include relevant bibliographic details) and the key bibliographic pages from books you consult. Take care with books as the full bibliographic details you will need are often located on two separate pages: title and author on one; and publisher, date and place of publication on another. All of this information also usually appears on a book's imprint page, which is located near the front of the book, ahead of the contents page(s). Have a look at those pages of this book for an example. **Reference management software** such as EndNote now allows relatively straightforward management of the bibliographic information you gather and helps to make it easier to prepare the **reference list**. Second, when you are writing an essay, be sure to insert **citations** as you go along. It is very difficult—and stressful—to come back to an essay and try to insert the correct references (Hodge 1994). Again, software packages like EndNote can be helpful with this.

Ensure you have an adequate number of sources

Markers will consider very carefully the quality of evidence you use in your work. You are expected to demonstrate that you have conducted extensive research appropriate to the topic and to the level of the course you are doing. For example, first-year essays might draw from secondary sources such as books, journals and the web, whereas third-year research essays might require library research, interviews, fieldwork and information derived from other primary information sources.

You will be expected to draw your evidence from, and to substantiate your claims using, *up-to-date*, *relevant* and *reputable* sources. Reputable sources might include **scholarly journals**, textbooks, high-quality websites (see Table 1.1), major newspapers (for example, the *Australian*, the *Age*, *Sydney Morning Herald*, *New Zealand Herald*, *Christchurch Press*, *Christian Science Monitor* and *New York Times*) and magazines (such as *Time* and *Far Eastern Economic Review*). Practise caution with all your sources, however. As noted earlier, simply because something is written in a journal, newspaper or on the web does not necessarily mean it is 'true'. Think critically about the trustworthiness of your sources—particularly if you are using material from the web. If you are in any doubt, ask your lecturer.

In much the same way as there is no answer to the question 'How long is a piece of string?', there is no specific number of sources you should consult for any particular kind or length of assignment. For instance, you cannot assume that twenty references is the right number for a first-year, 2000-word essay or that thirty is the correct number for a third-year, 4000-word research report. However, you should not rely on a small number of references or on references drawn too heavily from one source or type of information medium. Most assessors give some weight to the *number and range of references* you have used for your work. This is to ensure that you have established the soundness of your case by considering evidence from a broad range of possible sources. For example, the assessor might question the accuracy of an essay examining the consequences of hospital privatisation for Australian rural health care delivery if the essay were based largely on documents produced by the Liberal Party or by the Labor Party. Remember, too: most markers will expect you to have consulted sources other than free-access websites!

Make sure you acknowledge your sources adequately

Ideas, facts and quotations *must* be attributed to the source from which they were derived. Failure to acknowledge sources remains a common and potentially dangerous error in student essays (see Burkill & Abbey 2004).

Serious omissions may constitute **plagiarism** (see the notes on plagiarism in Chapter 10). **Acknowledgment** should be made using an appropriate system of referencing.

This business of acknowledging sources may seem to be a painful waste of time. But, like most things, there are reasons for it, as discussed in Box 1.7.

**Box 1.7
Reasons for acknowledging sources**

Citing sources fully and correctly is an important academic skill. Take the time to learn how it is done.

People writing in an academic environment acknowledge the work of others for three main reasons (Harris 2011):

- to attribute credit (and sometimes blame) for the acknowledged author's contribution to knowledge
- to avoid charges of plagiarism
- to allow readers the opportunity to pursue a line of inquiry should they be stimulated by something that has been cited. For example, in an essay on the economic geography of South Australian brewing, someone might write: 'Jones (2012, p. 16) observes that there is a great deal of money to be made by private investors by investing in the brewing business.' 'Aha,' you think, 'I can make some quick dollars here. All I need to do is find Jones's book and read about how this might be done.' Thanks to the appropriate acknowledgment of Jones's idea in the essay you have read, you can rush to the nearest library and track down Jones's words of wisdom on money-making. Lo and behold, instant millionaire!

Quite often, people new to essay writing leave acknowledgment of all references contained in a paragraph until the end of that paragraph. There they place a string of names, dates and page numbers. This is incorrect and annoying because the reader has no way of establishing which ideas, concepts or facts are being attributed to whom. References should be placed as close as possible to the ideas or illustrations to which they are connected.

If you are *quoting* someone *directly*, there are four golden rules to follow:

- *Reproduce the text exactly.* Spelling, capitalisation and paragraphing must mirror that of the original source. If a word is misspelt or if there is an error of fact, put the word '**sic**' in square brackets immediately after the error. This lets your reader know that the mistake was in the original source and that you have not misquoted. Your main text and the quotation should be grammatically consistent. This sometimes requires that you add or remove words. If you find it necessary to omit unnecessary words, use three full stops (...), known as an ellipsis, to show that you have deleted words from the original text. If you need to add words, put those you have added within square brackets. Make sure you do not change the original meaning of the text through your omissions or additions.

- When making a direct quotation of less than about thirty words you should *incorporate the quote* into your own text, indicating the beginning and end of the quote with single quotation marks. The in-text citation or numerical reference (see Chapter 10) is usually placed after the closing quotation marks. For example:

> He described Hispaniola and Tortuga as densely populated and 'completely cultivated like the countryside around Cordoba' (Colon 2006, p. 165).

> He described Hispaniola and Tortuga as densely populated and 'completely cultivated like the countryside around Cordoba'.[5]

If you are making a direct quotation of more than thirty words, you should not use quotation marks, but rather *indent and use single-spacing* as shown in the example here:

One historian makes the point clear:

> Although it included a wide range of human existence, New York was best known in its extremes, as a city capable of shedding the most brilliant light and casting the deepest shadows. Perhaps no place in the world asked such extremes of love and hate, often in the same person. (Spann 2001, p. 426)

A blank line immediately precedes and follows the quote.

- *Use quotations sparingly.* Only use a quotation when it outlines an idea or example so well that you cannot improve on it, or when it contains a major statement you must document.
- *Integrate the quotation into your text.* **Justify** its inclusion. Let the reader know what it means for your work.

See Chapter 10 for more details on acknowledging the works of others.

Make sure your in-text citation style is correct and consistent

Full and correct acknowledgment of the sources from which you have derived quotations, ideas and evidence is a fundamental part of the academic enterprise. Acknowledging the contribution of others to the essay you have written should not be difficult if you follow the instructions on referencing provided in Chapter 10.

Make sure your reference list is correctly presented

The most common—and most easily rectified—problems in essay writing emerge from incorrect acknowledgment of sources. Repeatedly in students' essays, referencing is done improperly and reference lists are formatted incorrectly. Many of those people who mark essays consider that problems with the relatively simple matter of citations reflect more serious shortcomings in the work they are reading. Consequently, it is advisable to follow carefully the instructions on acknowledging sources. If you do not understand how to refer to texts in an essay, see your lecturer, your tutor or a counsellor in your university's study skills centre. As noted earlier, you may also find it helpful to use reference management software to help ensure that you prepare a correctly laid-out reference list.

Before you submit an essay for assessment, be sure that all in-text citations have a corresponding entry in the list of *References Cited*. Further, in most cases, the list of *References Cited* should include *only* those references you have actually cited in the paper. This is usually contrasted with a **bibliography**, which comprises all the sources you have consulted.

Demonstrated level of individual scholarship

Essay markers want to know what you think and what you have learnt about a specific topic.

'Scholarship' is one of the most important and perhaps the least tangible of the qualities that make a good essay. Above all, the essay should clearly be a product of *your* mind and of *your* logical thought. The marker will react less than favourably to an essay that is merely a compilation of the work of other writers. In considering matters of scholarship, then, essay markers are searching for judicious use of reference material combined with *your individual insights*.

While scholarship requires that you draw from the work of other writers, you must do so with discretion. Use direct quotations sparingly (if you do use quotations, be sure to integrate them with the rest of your text) and keep paraphrasing to a minimum.

You might argue that novice status in the discipline means that you must rely heavily on other people's work. Obviously you might encounter problems when you are asked to write an essay on a subject that, until a few weeks ago, may have been quite foreign to you. But do not be misled into believing that in writing an essay you must produce some earth-shattering exposition on the topic you have been assigned. Instead, your assessor is looking for evidence that you have read on the subject, *interpreted* that reading, and set out appropriate evidence, based on that interpretation, which satisfactorily addresses the essay topic. Remember the advice from the beginning of the chapter. Your marker wants you to **explain** clearly what you think and what you have learnt about a specific topic.

Figure 1.1 Essay assessment schedule

Student Name:	Grade:	Assessed by:

The following is an itemised rating scale of various aspects of written assignment performance. Sections left blank are not relevant to the attached assignment. Some aspects are more important than others, so there is no formula connecting the scatter of ticks with the final grade for the assignment. Ticks in either of the two boxes left of centre mean that the statement is true to a greater (outer left) or lesser (inner left) extent. The same principle applies to the right-hand boxes. If you have any questions about the individual scales, comment, final grade or other aspects of this assignment, please see the assessor indicated above.

Quality of argument

The essay fully addresses the question	☐☐☐☐	The essay fails to address the question
Logically developed essay	☐☐☐☐	Writing rambles and lacks logical continuity
Writing well structured through introduction, body and conclusion	☐☐☐☐	Writing poorly structured, lacking introduction, cohesive paragraphing and/or conclusion
Material relevant to topic	☐☐☐☐	Much material is not relevant
Topic dealt with in depth	☐☐☐☐	Superficial treatment of topic

Quality of evidence

Essay well supported by evidence and examples	☐☐☐☐	Inadequate supporting evidence or examples
Accurate presentation of evidence and examples	☐☐☐☐	Much evidence incomplete or questionable
Effective use of figures and tables	☐☐☐☐	Figures and tables rarely used or not used when needed

Figure 1.1 *(cont.)*

Illustrations effectively presented and correctly cited				Illustrations poorly presented or incorrectly cited

Written expression and presentation

Fluent and succinct piece of writing				Clumsily written, verbose, repetitive
Grammatical sentences				Many ungrammatical sentences
Correct punctuation				Poor punctuation
Correct spelling throughout				Poor spelling
Legible, well set-out work				Untidy and difficult to read
Reasonable length				Too long/short

Sources/Referencing

Adequate number of references				Inadequate number of references
Adequate acknowledgment of sources				Inadequate acknowledgment of sources
Correct and consistent in-text referencing style				Incorrect and inconsistent in-text referencing style
Reference list correctly presented				Errors and inconsistencies in reference list

Demonstrated level of individual scholarship

High				Low

Assessor's comments

REFERENCES AND FURTHER READING

Allen, M. 2004, *Smart Thinking: Skills for Critical Understanding and Writing*, 2nd edn, Oxford University Press, Melbourne.

Anderson, J. & Poole, M. 2001, *Thesis and Assignment Writing*, 4th edn, Wiley, Brisbane.
    A comprehensive review of essay writing mechanics.

Barker, J. 2007, *Evaluating Web Pages Checklist,* viewed 20 May 2011, <www.lib. berkeley.edu/TeachingLib/.../webeval-QuestionsToAsk.pdf>.
    A very helpful page-long practical resource for those seeking detailed guidance on assessing the quality of web pages.

Barrett, H.M. 1982, *One Way to Write Anything*, Barnes and Noble, New York.
    Includes a detailed chapter on writing effective paragraphs.

Bate, D. & Sharpe, P. 1990, *Student Writer's Handbook*, Harcourt Brace Jovanovich, Marrickville, NSW.

Beck, S.E. 2009, *Evaluation Criteria*, viewed 6 May 2011, <http://lib.nmsu.edu/instruction/ evalcrit.html>.

Berger, A.A. 2008, *The Academic Writer's Toolkit. A User's Manual*, Left Coast Press, Walnut Creek, California.
    This book is aimed at an audience of academic scholars, but the advice set out in the first five chapters on matters like structure and style and moving from outline to draft is helpful.

Booth, V. 1993, *Communicating in Science: Writing a Scientific Paper and Speaking at Scientific Meetings*, 2nd edn, Cambridge University Press, Cambridge.
    Includes a very helpful review of pre-writing and writing strategies. Enjoyable reading.

Burdess, N. 1998, *Handbook of Student Skills*, 2nd edn, Prentice Hall, Sydney.

Burkill, S. & Abbey, C. 2004, 'Avoiding plagiarism', *Journal of Geography in Higher Education*, vol. 28, no. 3, pp. 439–46.

Clanchy, J. & Ballard, B. 1997, *Essay Writing for Students: A Practical Guide*, 3rd edn, Addison Wesley Longman, Melbourne.
    A best-selling book that covers the entire essay-writing process, including choosing a topic, taking notes, planning the answer, drafting and redrafting, and assessment. Well worth reading.

Davis, J. & Liss, R. 2006, *Effective Academic Writing 3. The Essay*, Oxford University Press, New York.
    Includes units on different kinds of essay, such as cause and effect, classification and reaction essays.

Dixon, T. 2004, *How to Get a First. The Essential Guide to Academic Success*, Routledge, London.

Includes a sometimes entertaining chapter on planning an essay that focuses on the importance of both thinking and planning before writing.

Fletcher, C. 1990, *Essay Clinic: A Structural Guide to Essay Writing*, Macmillan, South Melbourne.

This short book comprehensively outlines steps in the planning and construction of descriptive (word picture), narrative (storytelling), discursive (different viewpoints), expository (explanatory), analytical (dismantle and understand) and argumentative (persuasive) essays.

Friedman, S.F. & Steinberg, S. 1989, *Writing and Thinking in the Social Sciences*, Prentice-Hall, New York.

A valuable reference on all stages of the writing process.

Game, A. & Metcalfe, A. 2003, *The First Year Experience: Start, Stay and Succeed at Uni*, Federation Press, Leichhardt.

Includes a good chapter on writing that deals gently with the 'psychology' of writing as well as its mechanics.

Godwin, J. 2009, *Planning Your Essay*, Palgrave Macmillan, New York.

A useful pocket-size guide to essay planning.

Goldbort, R. 2006, *Writing for Science*, Yale University Press, New Haven, Connecticut.

This useful book covers a broad range of writing skills from laboratory notes, abstracts and memoranda to journal articles and grant proposals.

Greetham, B. 2008, *How to Write Better Essays*, 2nd edn, Palgrave Macmillan, Basingstoke, UK.

Includes useful material on writing effective paragraphs.

Harris, R.A. 2011, *Using Sources Effectively. Strengthening Your Writing and Avoiding Plagiarism*, 3rd edn, Pyrczak, Los Angeles.

An entertaining guide to effective referencing, means of avoiding unintentional plagiarism and improving your writing.

Henderson, E. & Moran, K. 2010, *The Empowered Writer. An Essential Guide to Writing, Reading, and Research*, Oxford University Press, Don Mills, Ontario.

A very thorough guide to high-quality academic writing, with detailed advice on a wide range of topics including essay types, essay design, paragraph construction and development, and information gathering.

Hodge, D. 1994, *Writing a good term paper*, course handout, Department of Geography, University of Washington, Seattle.

Lester, J.D. 1998, *Writing Research Papers*, 9th edn, Longman, New York.

Extensive coverage of the essay-writing process, moving from finding a topic to writing a proposal, doing library research, writing note cards and, eventually, writing the paper.

Lovell, D.W & Moore, R.D. 1992, *Essay Writing and Style Guide for Politics and the Social Sciences*, Australasian Political Studies Association, Canberra.

May, C.A. 2010, *Spotlight on Critical Skills in Essay Writing*, 2nd edn, Pearson Education Canada, Toronto.
A general guide to essay writing, featuring useful examples and exercises.

Moxley, J.M. 1992, *Publish Don't Perish: The Scholar's Guide to Academic Writing and Publishing*, Praeger, Westport, Connecticut.

Mullins, C. 1983, *A Guide to Writing and Publishing in the Social Behavioural Sciences*, Wiley, London.

Najar, R. & Riley, L. 2004, *Developing Academic Writing Skills*, MacMillan Languagehouse, Tokyo.

Northey, M. 2009, *Making Sense: A Student's Guide to Research and Writing*, 6th edn, Oxford, Toronto.
Several chapters of this successful book provide useful material on a variety of issues relating to essay writing.

Northey, M., Knight, D.B. & Draper, D. 2009, *Making Sense in Geography and Environmental Studies*, 4th edn, Oxford University Press, Toronto.
A fine guide to scholarly communication for students of Geography and Environmental Science.

Shields, M. 2010, *Essay Writing: A Student's Guide*, Sage, Los Angeles.
A comprehensive and well-received recent guide that deals with a broad range of issues involved in essay writing, including planning, information gathering and citing references.

Snooks & Co. 2002, *Style Manual for Authors, Editors and Printers*, 6th edn, John Wiley & Sons Australia, Canberra.

Taylor, G. 1989, *The Student's Writing Guide for the Arts and Social Sciences*, Cambridge University Press, Cambridge.
A detailed and successful review of the essay writing process.

Turner, K., Ireland, L., Krenus, B. & Pointon, L. 2008, *Essential Academic Skills*, Oxford University Press, South Melbourne.
Chapter 6 provides guidance on essay and reflective writing, and includes several useful activities intended to develop and hone relevant skills.

Warburton, N. 2007, *The Basics of Essay Writing*, Routledge, London.
A concise and comprehensive book that moves from cultivating good habits in essay writing to effective essay structure and tone, and then on to writing successful essay exams.

TWO

Writing a report

Research is the process of going up alleys to see if they are blind.

Marston Bates

The great tragedy of science—the slaying of a beautiful hypothesis by an ugly fact.

TH Huxley

Key topics

- Why write a report?
- What are report readers looking for?
- Components of a good report
- Writing a laboratory report

Your lecturer has assigned you a research project and, as if that was not difficult enough, you have been asked to report on your work—in writing. This chapter about writing research reports and laboratory reports is intended to help you complete that task. Indeed, by providing some guidance on how to write a particular kind of report, the following pages might also help you to undertake the research. Broadly speaking, reports can be distinguished from essays in that they typically present work which has yielded **primary data** in as dispassionate and impartial manner as possible, whereas essays usually draw from **secondary data** to support the author's considered opinions on a topic.

Why write a report?

There are at least three good reasons for learning to write good research reports. These range from the practical to the principled.

First, there is a *vocational claim*. Academic and professional writing often involves the communication of research findings. Indeed, as Montgomery (2003, pp. 138–9) notes, 'because of the tendency to outsource analysis and research these days, particularly in industry, the number and diversity of technical reports have gone through a burst of expansion'. Urban planners, market researchers, academics, environmental scientists and intelligence analysts can all expect to undertake research and to write associated reports in the course of their employment. Getting and maintaining employment in areas related to geography and the environment often requires the effective conduct and communication of research. This communication is usually to an audience that anticipates answers to a certain set of questions that must nearly always be answered, irrespective of the character of the project. Consequently, it is important to be familiar with the ways in which research results are customarily conveyed from one person to another (that is, the conventions of research communication).

Second, research reports and papers are a *fundamental and increasingly important building block of knowledge*. Each report is the final product of a process of inquiry such as a laboratory experiment or field research that helps us understand the ways in which the world works. Through the communication of research findings we contribute to the development of practically adequate understandings of the ways in which the world works. 'Practically adequate' means those understandings will not necessarily be absolutely and forever right. Instead, they work and make sense here and now. Some event or discovery may see them change tomorrow.

Third, there is a *moral responsibility to present our research honestly and accurately*. Through our research writing we help to forge understandings about the ways in which the world works. Representing the world to other people in ways that we understand is to play an enormously powerful role. To a degree, people entrust us with the creation of knowledge. Given that trust, our actions must be beyond reproach. In partial acknowledgment of that provision of trust, we must provide peers, colleagues and interested observers with accurate representations of our actions. Therein lies a critical role of research reports and a most important reason for writing them well.

What are report readers looking for?

Research and laboratory reports typically answer five classic investigative questions (Eisenberg 1992, p. 276), as shown in Box 2.1.

**Box 2.1
Five investigative questions**

- What did you do?
- Why did you do it?
- How did you do it?
- What did you find out?
- What do the findings mean?

The person reading or marking your report seeks clear and accurate answers to these questions. Because reports are sometimes long and complex, the reader will also appreciate some help in navigating their way through the document. You should make the report clear and easy to follow through easily understood language, a well-written introduction, suitable headings and subheadings, and, if appropriate, a comprehensive table of contents.

Some forms of report, especially laboratory reports, will answer the five investigative questions through a highly structured progression (for example, introduction, methods, results and discussion) written in a way that would allow another researcher to repeat the work. For example, an environmental scientist reviewing present-day salinity levels in Australia's Murray River, or a demographer conducting a statistical study on the use of contraceptive measures in New Zealand, is likely to conduct the study as impartially as possible and to record their research procedures in sufficient detail to allow someone else to reproduce the study. For such forms of inquiry, repetition is an important means of verifying results. (This notion of reproducibility or **replication** is discussed fully in Sayer 1992.)

In other forms of research, such as those involving qualitative research methods (for example, interviews, participant observation or textual analysis), results are confirmed in different ways. As a consequence, the research report may be written differently. It will usually answer the five questions identified above, but less emphasis will be given to the business of ensuring replicability. It is more important that qualitative research reports be written in a way that allows other people to confirm the reliability of your sources and to check your work against other related sources about the same or similar topics.

Consider, for example, the way a murder trial is conducted. The murder itself cannot be repeated to allow us to work out who the murderer was (replication). Instead, lawyers and police assemble evidence to reconstruct the crime as fairly and accurately as possible. This process is known as **corroboration**. Similarly, your report should be a fair and reasonable representation of events. (For fuller discussion of these and related matters, see Mansvelt & Berg 2010 and DeLyser & Pawson 2010.)

Geographers reporting on the social construction of an Australian city, or about gay men's perceptions of everyday places, will use different procedures and will write their research in different ways from coastal geomorphologists studying sand-grain size and longshore drift or economic geographers writing about the demographic characteristics of Australian country towns. Yet they will still usually answer the five basic investigative questions identified above, even though style and emphases may make the reports quite different in presentation.

The great diversity of research topics found within geography and environmental science means that you are likely to be asked to write research reports of different types throughout your degree program. Reflecting that potential diversity, this chapter provides an introduction to report writing *in general*. Some specific references are made to laboratory writing. It is very helpful to have available an example of a high-quality report that satisfies the requirements of your lecturer or whoever else 'commissioned' the report. You can use this as a general model of the kind of style you should be adopting or as a precise blueprint, depending on the expectations of your audience. So, if you are preparing a report for your lecturer, ask if they have written reports you may consult as examples. Alternatively, try your library.

Find 'model' reports to help guide you through your first report-writing efforts.

Components of a good report

It should be clear from the paragraphs above that although research and laboratory reports will usually answer Eisenberg's five investigative questions, there is no single correct research report style. The best way to organise a research report is determined by the type of research being carried out, the character and aims of the author, and the audience for whom the report is written. Accordingly, the following guidelines for report writing cannot offer you a recipe for a 'perfect' research report. Keys to a good report include well-executed research and the will and skills to communicate the results of your work effectively.

Having acknowledged that there is no single 'correct' report writing style, it is fair to say that over time a common pattern of report presentation has emerged. That pattern reflects a strategy for answering the five investigative questions identified above. Through repeated use, it is also a structure of presentation many readers will expect to see. If you are new to report writing, and unless you have been advised otherwise, it may be useful to follow the general pattern outlined below. If you are more experienced and believe there is a more effective way of communicating the results of your work, try out your own strategy. Remember, however, that you are guiding the reader through the work: you will have to let your audience know that you are doing something they might not expect.

Short research reports and laboratory reports generally comprise a minimum of seven sections that are outlined below. Long reports may add some or all of the extra materials listed. It is likely that most reports you are asked to write early in your university career will be short. By the later stages of your degree, in postgraduate work or in work environments, longer reports may be required.

Depending on the audience for your report, some of the headings listed in Table 2.1 may be eliminated, adapted or merged with one another. For instance, it may be appropriate in some instances to merge the discussion and recommendations sections in some long reports. Or there may be no need to include a letter of transmittal in a report written for your lecturer, although the inclusion of one might signal an added level of professionalism, as well as offering you experience in preparing such a letter.

Table 2.1 Contents of a report

Short report	Long report
• Title page	• Title page • Letter of transmittal (if appropriate)
• Abstract/executive summary	• Abstract/executive summary • Acknowledgments (sometimes placed after Discussion or immediately before References) • Table of contents
• Introduction (what you did and why)	• Introduction
• Materials and methods (how you did it)	• Materials and methods
• Results (what you found out)	• Results

Short report	Long report
• Discussion and conclusions (what the results mean)	• Discussion • Conclusion and recommendations • Appendices
• References	• References

Although these headings point to an order of report presentation, there is no need to write the sections in any particular sequence. Indeed, you may find it useful to follow Woodford's (in Booth 1993, p. 2) advice to label several sheets of paper or electronic files Title, Summary, Introduction, Methods, Results and so on and use these to record notes as you work through the project. Then begin your report by writing the easiest section (which is often the methods section).

The following pages outline the form and function of the common components of research reports. They also elaborate on some of the characteristics that contribute most significantly to effective research presentations in an academic setting. Those same characteristics form the basis of an assessment schedule for research reports and laboratory reports.

There is no need to write your report in the same order it will be presented.

Preliminary material

Title page

The best titles are usually short, accurate and attractive to potential readers. When you have finished writing your report, check that the title matches the results and discussion. An example of a functional and informative (though slightly bland) title is: 'Social consequences of homelessness for men in Adelaide, South Australia (2000–2012)'. This title lets the reader know the topic, place and time period. An example of a poor title on the same subject matter is: 'Men and homelessness'. If you can, try to devise a title that not only describes the research but also might encourage someone to read the report. For instance, despite its informative and comprehensive title, O'Loughlin and Witmer's (2011) paper, 'The Localized Geographies of Violence in the North Caucasus of Russia, 1999–2007', still raises questions that might encourage someone to read the work. For instance, just how did the researchers gather data on localised geographies of violence? At what scale were they working?

The title page of a report should also include:

• your name, position and organisational affiliation (or your name and student number)

- the name of the person and/or organisation to whom the report is being submitted (or the name of the lecturer and course for which the report was prepared)
- the date the report was issued (or submitted for assessment).

As the matters in parentheses suggest, you should modify these recommendations to suit the setting in which you find yourself.

Letter of transmittal

Reports are often commissioned by organisations such as government departments, private companies and charities. In such instances, a letter of transmittal is usually included in the final report. This letter personalises the report for the reader who commissioned (or asked for) the report and typically:

- explains the purpose of the letter; for example, 'Enclosed is the final report on wetland management issues in the Lake Ngaroto region that was commissioned by your organisation.'
- sets out the main finding of the report and any other vital issues likely to be relevant
- acknowledges any significant assistance received; for example, 'We are indebted to the Organisation of Lake Ngaroto Wetland Lovers for allowing us access to their extensive photographic records of post-1955 change in the lake.'
- offers thanks for the opportunity to conduct the research; for example, 'We would like to thank the Department of Conservation for engaging us to conduct this research.' (Mohan, McGregor & Strano 1992, p. 227)

Abstract/executive summary

Other than the title, this is the section of the report most likely to be read. It is important, therefore, to make it easy to understand. An **abstract** is a brief, coherent and concise statement, intelligible on its own, that typically provides concise answers to each of the five investigative questions outlined at the beginning of this chapter: What did you do? Why did you do it? How did you do it? What did you find out? What do the findings mean?

Many readers decide whether or not to read a report on the basis of its title and abstract; make sure they are clear and accurate.

Abstracts are short (usually 100–250 words) and are designed to be read by people who may not have the time to read the whole report. They are *not* written in the form of notes. All information contained in the abstract must be discussed within the main report. Do not write an abstract as if it is the alluring back-cover blurb of a mystery novel. Let your readers know what your research is about—do not leave them in suspense. Put the abstract at the beginning of your report, although it will usually be the last section you write.

Abstracts may be subdivided into two main types, *informative* and *indicative*, although some combine features of both.

An **informative abstract** typically summarises primary research, and offers a concise statement of details of the paper's content, including aims, methods, results and conclusions, as discussed in Box 2.2.

Box 2.2
Examples of informative abstracts

Example 1

Bell, H. (2011). 'Listing, "significance" and practised persuasion at Spa Green housing estate, London', *Social and Cultural Geography*, vol. 12, no. 3, pp. 223–42.

Abstract

This paper deals with the everyday, embodied performance of special architectural interest as enacted by residents of Spa Green—a modernist housing estate in north London designed by Lubetkin and Skinner of the Tecton practice, drawing on the innovative engineering of Ove Arup. Spa Green was designated as being of 'special interest'—and listed at grade II* in 1998—on the advice of English Heritage. Using a composite of data gathered by way of interview, video and photographic methods, the paper explores the ways in which this nomination as 'special' has reshaped the ordinary (everyday) lives of Spa Green residents. Since listing, changes have come both in who lives in Spa Green and how they do so. For example, some long-established residents have transformed how they live in and talk about the estate since its change in status. Similarly, the special status of the estate has attracted new residents who are self-conscious in their approach to—and consumption of—a modernist residence. For both groups, listing and the nomination of significance has transformed their ordinary, domestic lives; altering what kinds of knowledges they have about their homes and how they perform those knowledges. The paper looks specifically at these transformations in—and through—the seemingly modest features of the kitchen fittings, window dressing and bathroom/toilet configurations. Drawing on the art historian Keith Moxey's notion of 'persuasion', the paper argues that residents are moved to perform the special interest status of their homes, so engaging in a form of embodied persuasion.

Source: Reprinted by permission of the publisher Taylor & Francis Ltd. The full article is available at <www.tandfonline.com>.

Example 2

Lal, K.K. & Nunn, P.D. (2011). 'Holocene Sea Levels and Coastal Change, South-west Viti Levu Island, Fiji', *Australian Geographer*, vol. 42, pp. 41–51.

Abstract

For the first time, a sediment core spanning the entire Holocene has been analysed from Fiji. The 6m core was obtained from the floor of an ancient coastal lagoon (palaeolagoon) adjacent to Bourewa, the site of the earliest known human settlement in this island group. The basal sediments, just above bedrock, date from 11470 cal bp. A major transition occurs around 8000 cal bp where marine influences on palaeolagoon sedimentation increase sharply. Full shallow-water marine conditions are attained around 4630 cal bp and last until 3480 cal bp, after which there is a regressive phase. The results agree with the area-specific predictions of sea level in the ICE-4G model, particularly in the timing of the highstand. In addition, the results support the ideas (a) that early human colonisation of Fiji occurred during the late Holocene regression, (b) that the first inhabitants of Bourewa utilised both nearshore marine (reefal) and brackish lagoon food sources, and (c) that the abrupt human abandonment of the area around 2500 cal bp could have been prompted by a reduction in these resources driven largely by sea-level fall.

An **indicative abstract** outlines the contents of a paper, report or book, but does not recount specific details. It is commonly used to summarise particularly long reports and book chapters, as outlined in Box 2.3.

**Box 2.3
Examples
of indicative
abstracts**

Example 1

Lockwood, M., Davidson, J., Curtis, A., Stratford, E. & Griffith, R. 2009, 'Multi-level Environmental Governance: lessons from Australian natural resource management', *Australian Geographer*, vol. 40, no. 2, pp. 169–86.

Abstract

The region has become a significant scale of governance for the implementation of public policy, including natural resource management (NRM). A community-based regional NRM governance model has been adopted by the Australian government in partnership with Australian state and territory governments. There have been persuasive advocates of this approach both within community organisations and government. Proponents point to advantages such as the capacity to integrate across social, environmental and economic issues; improved investment efficiency; ability to establish appropriate power-sharing and partnership arrangements; better conversion of planning products into on-ground outcomes; and community learning and capacity building. However, concerns have also been raised in the academic literature regarding insufficient devolution of power, lack of downward accountability, exclusion of some stakeholders from decision making, and inadequate vertical and horizontal integration. We interviewed representatives from each of the governance levels (national, state, regional) to examine these concerns, and in doing so identify the strengths and challenges of the Australian experiment with devolved NRM governance. We synthesise the interview data with insights from the literature and make observations on the current state of Australian NRM governance. From this analysis, we identify lessons from the Australian experience to inform the development of multi-level environmental governance systems.

Example 2

Peace, A. 2011, 'Barossa dreaming: imagining place and constituting cuisine in contemporary Australia', *Anthropological Forum*, vol. 21, no. 1, pp. 23–42.

Abstract

Regional cuisines have become a prominent feature in the consumer landscape of modernised societies. This article describes how a regional cuisine is being socially constituted in the Barossa Valley, one of the most important wine-growing areas in Australia. Initially, I detail how small farmers, winegrowers and other entrepreneurs idealise the Barossa landscape and fabricate the heritage that is integral to the idea of a distinctive cuisine. This is followed by examination of how the notion of Barossa food as having distinctive qualities because of the artisanal ways it is produced is constantly being elaborated by the valley's small-scale enterprises. Lastly, I explore the contribution of wider influences to this cultural process, from the role played by an internationally recognised celebrity chef through to the recent arrival of the Slow Food movement. Local factors and global influences contribute to the social manufacture of the Barossa's regional cuisine, the overall appeal of which to middle-class consumers is as much cultural as it is culinary.

Abstracts usually comprise a single paragraph, although long abstracts may require more. They do not usually contain tables, figures or formulae and they should not discuss anything not covered in the paper or report.

All unfamiliar terms should be **defined**, as should **acronyms** (for example, NATO or NAFTA) and non-standard abbreviations (for example, dBA or pJ). Avoid referring to other works in an abstract. If you do refer to specific works or individuals, they must, of course, be included in the list of references associated with the full paper.

Usually, issues included or discussed in the main paper are presented in the present tense, whereas what the author did and thought is written in the past tense; for example, 'This report describes the nature of chemical weathering on ...'; 'It was discovered that ...'; 'Moreover, weathering had the effect of ...'; or 'The report concludes that ...'.

Acknowledgments

If you have received valuable assistance and support from some people or organisations in the preparation of the report, they should be acknowledged. As a general rule, thank those people who genuinely helped with aspects of the work, such as proofreading, preparing figures and tables, solving statistical or computing problems, or taking photos.

Table of contents

This should accurately and fully list *all* headings and subheadings used in the report with their associated page numbers. The table of contents occupies its own page and must be organised carefully with appropriate spacing. Make sure that the numbering system used in the table of contents is the same as that used in the body of the report. After the table of contents, and on separate pages, are a list of figures and a list of tables. Each of these lists contains, for each figure or table, its number, title and the page on which it is located. If your report uses many abbreviations and acronyms, provide a list of these too, but make sure that you also define each one fully when it first appears in the text. Page numbers in the contents and lists of tables and figures should be aligned vertically as part of the report's neat and orderly presentation.

Introduction—why did you do this study?

When you write your introduction, imagine that readers are unfamiliar with your work and that they really do not care about it. Let your audience know why this report is important and exactly what it is about (see Box 2.4), but do not include data or conclusions from your study. When readers know what you are going to discuss, they are better able to grasp the significance of the material you present in the remainder of your report.

The introduction of a report answers the following questions:

- What question is being asked? (If appropriate, state your **hypothesis**.)
- What do you hope to learn from this research?
- Why is this research important? (What is the social, personal and disciplinary significance of the work? This usually requires a literature review.)

> **Box 2.4**
> **Contents of a report's introduction**

When your readers have finished reading your introduction, they should know exactly what the study is about, what you hope to achieve from it, and why it is significant. If they have also been inspired to read the remainder of the report, so much the better!

Use the introduction to convince your audience to read the rest of your report.

Literature review

As part of, or soon after, your introduction you may need to write a **literature review** to provide the background to, and justification for, your research. The literature review provides the scholarly context—and perhaps the rationale—for the work being reported. It is sometimes presented as a separate part of the report, after the introduction and before the discussion of materials and methods. A literature review is a comprehensive, but pithy and critical, summary of publications and reports related to your research. It should discuss significant other works written in the area and make clear your assessment of those works. It is worth noting here that in this context 'critical' does not mean 'negative'. Instead it means scrutinising previous work from an analytical perspective, trying to take as dispassionate a view as possible while answering questions like: Do you agree with the author(s)? What are the work's strengths and weaknesses? What questions have the authors left unanswered? How does their work relate to that which your report discusses?

The literature review serves a number of functions (Deakin University Library 2011). It may help to:

- prevent you from 'reinventing the wheel' (that is, replicating the earlier work of others)
- identify gaps in the literature and potential research areas
- increase your breadth of knowledge in the field and highlight information, ideas and methods that might be relevant to your project
- identify other people working in the same area
- put your work into intellectual and practical perspective by identifying ways in which it may contribute to, fit in with, or differ from available work on the subject.

A good literature review makes clear the relationships between your work and that which has previously been done in the area.

Starting a literature review can be difficult. It is probably useful to identify the parent discipline(s) with which your research is associated (for example, geography or hydrology) and to then consult recent issues of the leading journals in the field to gather information on the topic area as broadly as it might be defined. Draw from the references in those articles to get some sense of the history of your research topic and to identify key authors, texts and articles. Gather those texts and read them. You may also find it helpful to consult the Social Sciences Citation Index (SSCI) or the Sciences Citation Index (SCI) to **trace** the intellectual genealogy of key references through time (speak to your librarian about this helpful technique). You may also wish to assemble a list of key words related to your topic and use electronic **databases** (for example, GEOBASE and Expanded Academic ASAP) to track down relevant articles.

Some advice set out in the previous chapter bears repeating here. First, keep a full record of the bibliographic details of all the references you use (for example, author, date of publication and article title). Second, be sure to insert citations as you write. Reference management software like EndNote can help with these tasks by, for example, allowing you to import bibliographic details directly from electronic databases. Once you have gathered and read relevant resources, you can begin writing the literature review. As a rule of thumb, a good literature review would normally discuss the truly significant books written in the field, notable books and articles produced on the broad subject in the past four or five years, and all available material on your specific research area. As noted above, the review should provide the reader with an understanding of the conceptual and disciplinary origins and significance of your study. It requires careful writing. As Reaburn (in Central Queensland University Library 2000) observes:

> Students will get a pile of articles and will regurgitate what article one said, what article two said. I can't emphasise enough, a well written literature review must evaluate all the literature, must speak generally, with general concepts they have been able to lift from all the articles, and they must be able to evaluate and critically analyse each one, then link and make a flow of ideas. Rather than separate little boxes, each box representing an article, make a flow of ideas, generalise and use specifics from one or two articles to back up a statement.

Your literature review should not be a string of **quotations** or a review of findings of other authors' work. Do not make the mistake of trying to list and summarise all material published in your area of work. Rather, organise the review into sections that present themes and trends related to your research.

Integrate all the little pieces of knowledge you have found during your reading into a coherent whole. Your completed literature review should be a critical analysis of earlier work set out in such a way that it becomes evident to the reader why your research work is being conducted.

Check to see that your literature review tells a coherent story about the development of research relevant to your topic.

You will probably find it helpful to look at examples of good literature reviews in your area of interest to gain a sense of how they are written. Short reviews are included as part of research papers in virtually every good academic journal. Have a look, for instance, at the articles by Lucas, Munroe and Pigozzi in a single issue of *The Professional Geographer*.

Lucas, S. 2004, 'The images used to "sell" and represent retirement communities', *The Professional Geographer*, vol. 56, no. 4, pp. 449–59. Lucas confines her review to pages 450–3 in particular.

Munroe, D.K., Southworth, J. & Tucker, C.M. 2004, 'Modeling spatially and temporally complex land-cover change: the case of Western Honduras', *The Professional Geographer*, vol. 56, no. 4, pp. 544–59. The literature review in this paper is brief and confined largely to page 545.

Pigozzi, B.W. 2004, 'A hierarchy of spatial marginality through spatial filtering', *The Professional Geographer*, vol. 56, no. 4, pp. 460–70. [Pigozzi's review is set out in the section entitled 'Introduction and context'.]

Materials and methods—how did you do this study?

You should provide a precise and concise account of the materials and methods used to conduct the study, and why you chose them. Let your reader know exactly how you did the study and where you got your data. A good description of materials and methods should enable readers to replicate the investigative procedure even if they have no source of information about your study other than your report. In qualitative studies, however, replication of procedure is unlikely to lead to identical results. Instead, the outcome may be results that corroborate, substantiate or, indeed, refute those achieved in the initial study.

If your reader had no source of information other than your report, could they repeat your study?

Depending on the specific character of the research, the methods section of a report comprises up to three parts (Dane 1990, pp. 219–21), which may be written as a single section or presented under separate subheadings: sampling, apparatus and procedure.

Sampling/subjects

An important part of the materials and methods section of a report is a statement of *how* and *why* you chose a particular place, group of people or object to be the focus of your study. For example, if your research concerns people's fears and the implications those fears have for the use of urban space, why have you chosen to confine your study to a specific suburb of one Australian city? Having limited the study to that location, why and how did you choose a small group of people to speak to from the much larger total local population? Alternatively, in an examination of avalanche hazards in New Zealand's South Island high country, why and on what bases did you limit your study to those risks associated with one popular ski area?

Explain and justify your sampling and case selection decisions.

In the sampling or subjects section of your report, your reader will appreciate answers to the following questions:

- *Whom or what* specific group, place or object have you chosen to study? You may have already stated in the introduction that you were exploring the attitudes of Papua New Guinean women to birth control, but you now need to identify the specific group and number of women you are going to interview or to whom you will administer questionnaires (for example, 500 urban-dwelling women of child-bearing age). Or, in your study about supernatural explanations of unusual landscape features, you might have determined that you will limit your study to Incan constructions in Peru within a 100-kilometre radius of the historically important town of Cuzco.
- *Why* did you make that choice? Why did you limit the study to 500 Papua New Guinean women of child-bearing age and not to a smaller group of rural-dwelling women? Or, in our other example, why Peru and not Easter Island? Why 100 kilometres? Why Cuzco and not Machu Picchu or Aguas Calientes?
- *How* did you select the unit(s) of study? That is to say, what specific sampling technique did you employ (for example, snowball, simple random, typical case, cluster area or random traverse)? There is no need to go into great detail about the technique, such as describing any computer programs used in your sampling, unless the procedure was unusual.
- *What* are the limitations and shortcomings of the data or sources?

Don't hesitate to illustrate.

Of course, if you are reporting a field study, a general description of the study site is needed. Do not forget to include a map: it may save you a great deal of writing and will almost certainly provide your reader with a clearer sense of the place you are describing than might the proverbial thousand words. Photographs also may be helpful.

Apparatus or materials

Provide a brief description of any special equipment or materials used in your study. For example, briefly describe any experimental equipment or questionnaires used in your work. In some more advanced studies you may also be expected to include the name and address of equipment manufacturers in **parentheses**. Do not hesitate to use figures and plates in your description of apparatus.

Procedure

This section contains specific *details* about how the data were collected, about response levels, and about the methods used to interpret the findings. For example, if your study was based on a questionnaire survey or experimental procedure, tell your readers about the process of questionnaire administration, or about the experiment, in enough detail to allow them to replicate your procedures. What statistical tests did you decide to use? It is important to *justify* your selection of data collection and statistical procedures in this section. Why did you choose one method over others? Give references to support your selection.

What are the advantages and disadvantages of the procedure you selected and how did you overcome any problems you encountered? You might consider it more appropriate to confine discussion of this last question to the discussion section of your report.

Where appropriate (for example, in research involving human or animal subjects) your discussion of procedural issues should also indicate how the work satisfied relevant ethical guidelines (for a full discussion of research **ethics**, see Israel & Hay 2006).

Results—what did you find out?

The results section of a research report is typically a dispassionate, factual account of findings. It outlines what occurred or what you observed. State clearly whether any hypotheses you made can be accepted or rejected, but in general you should not discuss the significance of those results here. That will be covered by the next section of the report. For example, you may have conducted a study that suggests that all koalas on a small offshore island will starve to death unless something is done to control their population. You would save your discussion of ways to resolve that problem for a later section of the report.

Although it is not customary to present conclusions and interpretations in the results section, in some qualitative and laboratory reports it is considered appropriate to combine the results with an interpretive discussion. If you have any doubt about what is appropriate, ask your lecturer.

A key to effective presentation of results is to make them as comprehensible to your readers as possible. To this end, it may be appropriate to begin your

Results are important, but save time, space and energy to interpret them.

discussion of results with a brief overview of the material that is to follow before elaborating. You might also consider presenting your results in the chronological order in which you discovered them.

Use maps, tables, figures and written statements creatively to summarise and convey key information emerging from the study. If you have provided results in figures and tables, do not repeat all the data in the text. Emphasise only the most important observations. You should place tables and figures close to the text in which they are mentioned without interrupting the flow of the text. However, if you have particularly detailed and lengthy data lists or figures that supplement the report's content, these may be better placed in an appendix.

The results section will often contain a series of subheadings. These usually reflect subdivisions within the material being discussed, but sometimes reflect matters of method. In general, however, try to avoid splitting up the results section on the basis of methods, since it may suggest that you are 'allowing the methods rather than the issues to shape the problem' (Hodge 1994, p. 2).

If you have not already done so in the report, the results section is also an appropriate place in which to identify the limits of your data.

Discussion and conclusion—what do the findings mean?

> I am appalled by ... papers that describe most minutely what experiments were done, and how, but with no hint of why, or what they mean. Cast thy data upon the waters, the authors seem to think, and they will come back interpreted.
>
> Woodford (1967, p. 744)

The **discussion** is the heart of the report. Perhaps not surprisingly, it is also the part that is most difficult to write and—after the title, abstract and introduction—the section most likely to be read thoroughly by your audience. Readers and assessors will be looking to see if your work has achieved its stated objectives. So, take particular care when you are writing this part of your report.

The discussion has two fundamental aims:

- to explain the results of your study. Why do you think the patterns—or lack of patterns—emerged?
- to explore the significance of the study's findings. What do the findings mean? What new and important matters have been raised? Compare

your results with trends described in the literature and with theoretical behaviour. Embed your findings in their larger academic, social and environmental contexts. Make explicit the ways in which your work fits in with studies conducted by other people and the degree to which it might have broader importance.

Use the discussion to explain results and explore significance.

Hodge (1994, p. 3) makes the point:

> Remember that research should never stand alone. It has its foundations in the work of others and, similarly, it should be part of what others do in future. Help the reader make those connections.

The concluding sections of the report might also offer suggestions about improvements or variations to the investigative procedure that could be useful for further work in the field. Where do we go from here? Are there other methods or data sets that should be explored? Has the study raised new sets of questions? (Hodge 1994, p. 3) Any thoughts you have must be *justified*. Many students find it easy to offer suggestions for change, but few are able to support their views adequately.

Recommendations

If your report has led you to a position where it is appropriate to suggest particular courses of action or solutions to problems, you may wish to add a recommendations section. This could be included within the conclusion or given a separate section. In some reports, recommendations are placed at the front, following the title page. Recommendations should be based on material covered in the report (Mohan, McGregor & Strano 1992, p. 228).

Appendices

Material that is not essential to the report's main argument and is too long or too detailed to be included in the main body of the report is placed in an appendix at the end of the report. For example, you might include a copy of the questionnaire you used, background information on your study area or pertinent data that is too detailed for inclusion in the main text. A separate appendix should be used for each type of material and should be labelled clearly (for example, Appendix 1, 2, 3 … or Appendix A, B, C, D …). However, your appendix should *not* be a place to put *everything* you collected in relation to your research but for which there was no place in your report. Appendices are usually located after the conclusions, but before the references.

Don't use appendices as a dumping ground for data.

References

For information on citing references in a research report, see Chapter 10.

Written expression and presentation

Language of the report

Some audiences reading research and laboratory reports still expect the report to be written in 'objective', dispassionate, **third-person narrative** style (for example, 'it was decided' rather than 'I decided'). Consider that expectation when writing your report. If you choose to write in **first-person narrative** style, which reflects the social creation of knowledge, some audiences may be distracted and unconvinced by your apparent 'personal bias'. Whatever choice you make, remember that simply writing a report in the superficially 'objective' third person does not render it any more accurate than a report written in the first person! (For a lengthy discussion of this, see Mansvelt and Berg 2010.) If you have any questions or concerns about the style of language you should use in your report, ask your lecturer.

Another matter of language that warrants attention is the use of **jargon**. The word 'jargon' has two popular applications. More commonly, jargon refers to technical terms used inappropriately or when clearer terms would suffice. More accurately, it means words or a mode of language intelligible only to a group of experts in a particular field of study (Friedman & Steinberg 1989, p. 30). There will be occasions in report writing when you will find it necessary to use jargon in the second sense of the word. Using topic-specific terms in the correct context can yield effective and efficient communication, as well as demonstrating to you assessor your depth of understanding of the subject material. However, you should never be guilty of using jargon in the more common, first sense of the term. Remember, you are writing to communicate ideas to the intended audience as clearly as possible. Use the language that allows you to do that. KISS (that is, 'Keep It Simple, Stupid') your audience. For a more detailed discussion of jargon, see Chapter 10.

Make sure your report is coherent and formatted consistently.

Two final points include checking that your report's text and figures 'move from the general to the specific ... for individual sections and for the report as a whole' (Montgomery 2003, pp. 143–4) and ensuring that your report—despite its many sections—reads as a single, integrated document. This is an especially important issue if your report is the output of a group project and sections have been written by different people or teams. It may involve the sometimes lengthy tasks of rewriting some sections, reordering elements of the report and ensuring all references, figures and tables are formatted uniformly.

Presentation

Be sure that your report is set out in an attractive and easily understood style. Care in presentation suggests care in preparation. *Care* is to be emphasised here, not gaudiness and decoration. People tend to be suspicious of overly 'decorated' reports, and in a professional environment, such as consulting, may also question the costs of production. Get the fundamental matters straight. For instance:

- Use the same-sized paper throughout (however, there may be some occasions when the use of same-sized sheets is impractical, such as when you are using maps).
- Number all pages.
- Ensure the different parts of your report stand out clearly.
- Use lots of white space—but do be judicious about your use of paper.
- Use SI (*Système International*) units (that is, metric) in describing measures.
- Use a clear and consistent hierarchy of headings. Unlike those used in essays, headings in reports typically are numbered. An example is set out in Box 2.5.

1 Introduction
   1.1 Background
   1.2 Aims
   1.3 Objectives
2 Methods
   2.1 The questionnaire
   2.2 Sample group
3 Results
   3.1 Response rate
   3.2 Findings
      3.2.1 Who is fearful in urban space?
      3.2.2 Precautionary strategies taken to avoid perceived threats
      3.2.3 Residents' views on means to reduce levels of fear
4 Conclusions
5 Recommendations
   5.1 Public solutions to fear of violence in urban space
   5.2 Individuals' solutions
6 Appendices
   i The questionnaire
   ii Tabulated responses
7 References

**Box 2.5
Example of heading hierarchies in a report**

Writing a laboratory report

A laboratory report is a particular form of research report. Hence, the preceding advice is applicable. Typically, a laboratory report dispassionately and accurately recounts experimental research procedures and results. A good report is written so that another student or researcher could repeat the experiment in exactly the same way as you did (assuming, of course, that you employed correct procedures) and could compare their results with yours. You must be both meticulous in outlining methods and accurate in your presentation of results. Meticulous does not mean tedious. Try not to overdo the detail. In the context of the specific experiment that you are doing, record those things that are important. What did you need to know to do the experiment? Let your reader know that. One could almost argue that the best 'Materials and methods' sections in laboratory reports ought to be written by novices who are less likely than more experienced researchers to make assumptions about readers' understanding of experimental methods.

By convention, laboratory reports follow the order of research report presentation outlined earlier in this chapter; that is:

- Title page
- Abstract
- Introduction
- Materials and methods
- Results
- Discussion
- Appendices
- References

Depending on the specific nature of your experiment, your lecturer or laboratory supervisor may not require all these sections. For example, if your report is simply an account of work you undertook during a single laboratory class, it is possible that there will be no need for you to include an abstract, references and appendices. However, your instructor may be impressed if you take the care and attention to relate your day's laboratory work to appropriate reference or lecture material. Clearly, though, if your laboratory work is conducted over a period longer than a single class, you will have the opportunity to consult relevant reference materials. Figure 2.1 shows aspects of a report that your assessor might use to mark your work. The detailed items offer a summary guide to the kinds of matters you might attend to when writing your report, as well as the basis for the provision of feedback.

Figure 2.1 Research report assessment schedule

Student Name:	Grade:	Assessed by:

The following is an itemised rating scale of various aspects of research report performance. Sections left blank are not relevant to the attached assignment. Some aspects are more important than others, so there is no formula connecting the scatter of ticks with the final grade for the assignment. Ticks in either of the two boxes left of centre mean that the statement is true to a greater (outer left) or lesser (inner left) extent. The same principle is applied to the right-hand boxes. If you have any questions about the individual scales, comment, final grade or other aspects of this assignment, please see the assessor indicated above.

Purpose and significance

Statement of problem or purpose is clear and unambiguous					Statement of problem or purpose is unclear or ambiguous
Research objectives outlined precisely					Research objectives unclear
Disciplinary, social and personal significance of the research problem made clear					Problem not set in context
Documentation fully outlines the evolution of the research problem from previous findings					No reference to earlier works or incorrect references

Description of method

Most appropriate research method selected					Research method selected is inappropriate
Sample, cases or study area appropriate to purpose of inquiry					Sample, cases or site unsuitable
Complete description of study method					Inadequate description of study method

Figure 2.1 *(cont.)*

Quality of results

Evidence of extensive primary research	☐	☐	☐	☐	Little or no evidence of primary research
Limitations of sources made clear	☐	☐	☐	☐	Inappropriate sources accepted without question
Relevant results presented in appropriate level of detail	☐	☐	☐	☐	Relevant results omitted or suppressed

Discussion & interpretation

No errors of interpretation (e.g. logic or calculation) detected	☐	☐	☐	☐	Many errors of interpretation
Limitations of findings made clear	☐	☐	☐	☐	Limitations of findings not identified
Discussion connects findings with relevant literature	☐	☐	☐	☐	No connection between findings and other works

Conclusions

Significance of findings made clear	☐	☐	☐	☐	Little or no significance identified
Conclusions based on evidence presented	☐	☐	☐	☐	Little or no connection between evidence and conclusions
Stated purpose of research achieved	☐	☐	☐	☐	Little or no contribution to solution of problem or achievement of purpose

Use of supplementary material

Effective use of figures and tables	☐	☐	☐	☐	Illustrative material not used when needed or not discussed in text
Illustrations presented correctly	☐	☐	☐	☐	Illustrations poorly presented

Detailed statistical analyses and tables placed in appendices				Excessively detailed findings in text

Written expression and presentation

Document follows assigned report format conventions				Little or no adherence to report presentation
Clearly and correctly written				Poor written expression
Report carefully produced				Sloppy presentation

Sources/referencing

Adequate number of sources				Inadequate number of sources
Adequate acknowledgment of sources				Inadequate acknowledgment of sources
Correct and consistent in-text referencing style				Incorrect or inconsistent referencing style
Reference list correctly presented				Errors and inconsistencies in reference list

Assessor's general comments

REFERENCES AND FURTHER READING

Beer, D.F. (ed.) 2003, *Writing and Speaking in the Technology Professions: A Practical Guide*, 2nd edn, Wiley-IEEE Press, New York.

This valuable edited collection comprises over 60 short papers on a wide variety of communication skills, including technical report writing. Other topics include oral presentations, running meetings, writing resumes, preparing illustrations and writing proposals.

Blicq, R.S. 1987, *Writing Reports to Get Results: Guidelines for the Computer Age*, IEEE Press, New York.

This book, of more than 200 pages, offers a serious, no-nonsense review of different report types and how to go about compiling them. Numerous examples are provided. Do not be misled by the volume's publishers, the Institute of Electrical and Electronics Engineers: this book is useful to students and practitioners in other fields.

Booth, V. 1993, *Communicating in Science. Writing a Scientific Paper and Speaking at Scientific Meetings*, 2nd edn, Cambridge University Press, Cambridge.

Chapter 1 of Booth's readable and brief book offers helpful advice on writing a scientific paper. Particular attention is devoted to the mechanics and detail of scientific presentation.

Brower, J.E., Zar, J.H. & von Ende, C.N. 1990, *Field and Laboratory Methods for General Ecology*, 3rd edn, WC Brown, New York.

Cargill, M. & O'Connor, P. 2009, *Writing Scientific Research Articles: Strategy and Steps*, Wiley-Blackwell, Chichester, UK.

Although this book is written for more senior scholars, it offers accessible, detailed and valuable advice on matters such as when and how to write each section of a report, how to write a compelling introduction, and structural issues to consider when writing a discussion section.

Central Queensland University Library 2000, *Why do a Literature Review?*, viewed 27 March 2001, <http//www.library.cqu.edu.au/litreviewpages/why.htm>.

Dane, F.C. 1990, *Research Methods*, Brooks/Cole, Pacific Grove, California.

Day, R.A. & Gastel, B. 2006, *How to Write and Publish a Scientific Paper*, Greenwood Press, Westport, Connecticut.

A comprehensive book with many chapters devoted to writing the various sections of a research report.

Deakin University Library 2011, *The Literature Review*, viewed 10 May 2011, <www.deakin.edu.au/library/findout/research/litrev.php>.

DeLyser, D. & Pawson, E. 2010, 'From personal to public: communicating qualitative research for public consumption', in *Qualitative Research Methods in Human Geography*, 3rd edn, ed. I. Hay, Oxford University Press, Melbourne, pp. 356–66.

Eisenberg, A. 1992, *Effective Technical Communication*, 2nd edn, McGraw-Hill, New York.

Friedman, S.F. & Steinberg, S. 1989, *Writing and Thinking in the Social Sciences*, Prentice-Hall, New York.

Chapter 3 includes useful discussions of the role of writing in the research process and the three components of the rhetorical stance: subject, audience and voice. Take care, however, for in their consideration of voice, the authors imply that sufficient evidence and emotive language are mutually exclusive. Clearly this is incorrect. There is also a helpful review of the appropriate and inappropriate uses of jargon in technical writing.

Hodge, D. 1994, 'Guidelines for Professional Reports', course handout for GEOG 426, Department of Geography, University of Washington, Seattle.

Israel, M. & Hay, I. 2006, *Research Ethics for Social Scientists. Between Ethical Conduct and Regulatory Compliance*, Sage, London.

Lucas, S. 2004, 'The images used to "sell" and represent retirement communities', *The Professional Geographer*, vol. 56, no. 4, pp. 449–59.

Mansvelt, J. & Berg, L. 2010, 'Writing qualitative geographies, constructing geographical knowledges', in *Qualitative Research Methods in Human Geography*, 3rd edn, ed. I. Hay, Oxford University Press, Melbourne, pp. 333–55.

Mohan, T., McGregor, H. & Strano, Z. 1992, *Communicating! Theory and Practice*, 3rd edn, Harcourt Brace, Sydney.
Chapter 9 offers an overview of report types, functions, format and style.

Montello, D.R. & Sutton, P.C. 2006, *An Introduction to Scientific Research Methods in Geography*, Sage, Thousand Oaks.
A comprehensive review of scientific research methods, with chapters on communication and research ethics.

Montgomery, S.L. 2003, *The Chicago Guide to Communicating Science*, University of Chicago Press, Chicago.
Chapter 10, on technical reports, is a very helpful discussion on the context within which reports are written and the form they should take.

Munroe, D.K., Southworth, J. & Tucker, C.M. 2004, 'Modeling spatially and temporally complex land-cover change: the case of Western Honduras', *The Professional Geographer*, vol. 56, no. 4, pp. 544–59.

Oliver, P. 2008, *Writing Your Thesis*, 2nd edn, Sage, Los Angeles.
As the title suggests, this book is intended for Honours and postgraduate students. However, much of the advice on matters such as writing a literature, methodology and data analysis has much broader application.

O'Loughlin, J. & Witmer, F.D.W. 2011, 'The Localized Geographies of Violence in the North Caucasus of Russia, 1999–2007', *Annals of the Association of American Geographers*, vol. 101, no. 1, pp. 178–201.

Pigozzi, B.W. 2004, 'A hierarchy of spatial marginality through spatial filtering', *The Professional Geographer*, vol. 56, no. 4, pp. 460–70.

Pyrczak, F. & Bruce, R.R. 2011, *Writing Empirical Research Reports: A Basic Guide for Students of the Social and Behavioral Sciences*, 7th edn, Pyrczak Publishing, Glendale, California.

An easy-to-follow guide offering advice and examples.

Sayer, A. 1992, *Method in Social Science. A Realist Approach*, 2nd edn, Routledge, London.

Sayer's influential book includes a useful chapter entitled 'Problems of explanation and the aims of social science', which introduces differences between 'intensive' (essentially qualitative) and 'extensive' (essentially quantitative) research and their implications for the process of research and the communication of results. However, this chapter is challenging reading.

Thody, A. 2006, *Writing and Presenting Research*, Sage, London.

This volume covers presentation of research results through various media; for example, theses, chapters, books, reports and articles in academic, professional or general media (such as newspapers).

University of California—Santa Cruz University Library 2011, *Write a Literature Review*, viewed 11 May 2011, <http://library.ucsc.edu/print/help/howto/write-a-literature-review>.

A very helpful practical resource for those seeking additional guidance on writing a literature review.

University of Wisconsin—Madison Writing Centre 2011, *Learn How to Write a Review of Literature*, viewed 11 May 2011, <http://writing.wisc.edu/Handbook/ReviewofLiterature.html>.

Another very helpful practical resource for those seeking additional guidance on writing a literature review.

Woodford, F.P. 1967, 'Sounder thinking through clearer writing', *Science*, vol. 156, no. 3776, p. 744.

Zeegers, P., Deller-Evans, K., Egege, S. & Klinger, C. 2008, *Essential Skills for Science and Technology*, Oxford University Press, South Melbourne.

Chapters 13 and 14 provide a helpful supplementary overview of writing research and laboratory reports.

THREE

Writing an annotated bibliography, summary or review

I was so long writing my review that I never got around to reading the book.

Groucho Marx

Some books are to be tasted, others to be swallowed and some few to be chewed and digested.

Francis Bacon (1625)

Key topics

- Preparing an annotated bibliography
- Writing a summary or precis
- Writing a good review

Your academic endeavours will often require you to summarise and make sense of the works of other people. This chapter provides some advice on writing annotated bibliographies, summaries (or precis) and reviews of books, articles and websites—exercises that specifically require you to interpret and abridge longer pieces of work in a comprehensible fashion. It also sets out the criteria that readers and assessors usually consider in evaluating this kind of work. You will find assessment criteria for different types of summary assignments in Figures 3.1 and 3.2.

Preparing an annotated bibliography

An **annotated bibliography** is a list of reference materials, such as books, articles and websites, in which you provide author, title and publication details for each item (as in a **bibliography**), together with a short review of that item. The review, which **summarises** and sometimes **critiques** the source material, is typically up to 150 words long. Annotated bibliographies are customarily set out with the items in alphabetical order (by author surname). However, some annotated bibliographies are written in the form of a short essay that quickly and concisely offers the same bibliographic material and critique, but in a more literary style than an annotated list.

Box 3.1 sets out four examples of items from annotated bibliographies. The first three employ the common review approach, while the fourth demonstrates the short essay approach.

**Box 3.1
Examples of annotated bibliographies**

Example 1

Northcott, M.S. 2007. *A Moral Climate: The Ethics of Global Warming*, Darton, Longman and Todd, London.

Response to the challenge of global warming requires learning to put the common good ahead of selfish interests, weaving together the physical climate and the moral climate. Relieving climate change opens opportunities for solving other problems: world poverty, the rich–poor divide, the overuse of resources, and the appreciation and conservation of human creation.

Source: Callicoat and Frodeman (2008).

Example 2

Sturgeon, N. 2009, *Environmentalism in Popular Culture: Gender, Race, Sexuality, and the Politics of the Natural*, University of Arizona Press, Tucson.

This is the first book to employ a global feminist environmental justice analysis to focus on how racial inequality, gendered patterns of work, and heteronormative ideas about the family relate to environmental questions. Beginning in the late 1980s and moving to the present day, Sturgeon unpacks a variety of cultural tropes, including ideas about Mother Nature, the purity of the natural, and the allegedly close relationships of indigenous people with the natural world. She investigates the persistence of the "myth of the frontier" and its extension to the frontier of space exploration. She ponders the popularity (and occasional controversy) of penguins (and penguin family values) and questions assumptions about human warfare as "natural."

Sturgeon illustrates the myriad and insidious ways in which American popular culture depicts social inequities as "natural" and how our images of "nature" interfere with creating solutions to environmental problems that are just and fair for all. Why is it, she wonders, that environmentalist messages in popular culture so often "naturalize" themes of heroic male violence, suburban nuclear family structures, and US dominance in the world? And what do these patterns of thought mean for how we envision environmental solutions, like "green" businesses, recycling programs, and the protection of threatened species?

Source: Reed (2011).

Example 3

Bullard, R.D. 1990, *Dumping in Dixie: Race, Class, and Environmental Equity*, Westview Press, Boulder, CO.

Dumping in Dixie is an in-depth study of environmental racism in black communities in the South. Bullard explores the barriers to environmental and social justice experienced by blacks and the factors that contribute to the conflicts, disparities, and the resultant growing militancy. He provides case studies of strategies used by grassroots groups who wanted to take back their neighborhoods in Houston's Northwood Manor neighborhood; West Dallas, Texas; Institute, West Virginia; Alsen, Louisiana; and Emelle-Sumter County, Alabama.

In these predominantly black communities, grassroots organizing was carried out to protest against landfills, incinerators, toxic waste, chemical industries, salvage yards, and garbage dumps. Strategies included demonstrations, public hearings, lawsuits, the election of supporters to state and local offices, meetings with company representatives, and other approaches designed to bring public awareness and accountability. Bullard offers action strategies and recommendations for greater mobilization and consensus building for the ensuing environmental equity struggles of the 1990s.

Source: Weintraub (1994).

Example 4

I know of no geographical research on the condom, and virtually nothing on the geography of sexual relations, although the fine and pioneering work by a geographer:

R. Symanski (1981, *The Immoral Landscape: Female Prostitution in Western Societies*, Butterworths, Toronto)

is increasingly referenced by other human scientists. Numerous short reports on condom use and propagation are given in almost all issues of *World–AIDS*, while many articles and reports in the 'AIDS Monitor' of *New Scientist* deal with condom use.

Source: Gould (1993, p. 215).

A good place to see examples of annotated bibliographies is the web. Some of these include, for example, Hadden and Minson's (2010) review of Haiti's geography, geology and earth science, Cottingham, Healey and Gravestock's (2002) collection on fieldwork in geography and environmental sciences, and Reed's (2011) work on environmental justice.

What is the purpose of an annotated bibliography?

Annotated bibliographies provide people in a particular field of inquiry with some commentary on books, articles, websites and other resources available in that field. They discuss the content, relevance and quality of the material reviewed (Engle, Blumenthal & Cosgrave 2011). Annotated bibliographies may also introduce a newcomer to a body of work through an insightful review of available material. Your instructor might ask you to write an annotated bibliography to make you familiar with some of the literature in your discipline area.

What is the reader of an annotated bibliography looking for?

Although the content may vary depending on the purpose of the list, readers of annotated bibliographies typically expect to see three sets of information (as discussed in Box 3.2), although many annotated bibliographies exclude the element of critique.

**Box 3.2
Key information in an annotated bibliography**

Details: full bibliographic details (see Chapter 10 for details).

Summary: a clear indication of the content (and argument) of the piece. Consider including, for example, material on the author's aim, their intended audience, their claim to authority and key arguments they use to support their points.

Critique: critical comment on the merits and weaknesses of the publication or on its contribution to the field of study. Some of the things you also might consider evaluating are: appropriateness of the article to its intended audience; whether it is up to date; and, crucially, its engagement with other important literature in the field.

Writing a summary or precis

Summaries (sometimes called **precis**) restate the essential contents of a piece of writing in a much more limited (and usually specified) number of words than the original text. Abbreviation of the text is accomplished by presenting the main ideas in alternative wording, leaving out most examples and minor points. Your precis must accurately *re-present* the text in condensed form; that is, it should be a scaled-down version of the original text. Unlike a review, a precis does *not* interpret issues raised. It is *not* evaluative. There is no need for you to provide your reaction to the ideas of the author.

A good precis spares the details and focuses on key matters.

Brevity and clarity are critical ingredients of a summary or precis. Let your reader or assessor know, in as few words as possible, what the summarised text is about. Do not prepare a precis that is as long as the article itself. Imagine your audience sitting opposite you, their eyes about to glaze over with boredom. Spare them the details. Give them enough information to understand what the text is about, but not so much that they might just as well have read the original.

This leads us to one of the most common shortcomings of a summary or a precis: the failure to say what the original author's *main argument* is. Imagine someone has asked you to tell them about a movie you have seen recently. One of the most important things they will want to know, and in relatively few words, will be *essential details of the plot*. For example, readers probably do not want to know the names of Cinderella's evil sisters, the colour of her ball gown or the temperament of the horses. Instead, they want to know the essence of the story that brought Cinderella together with her Prince Charming.

Once you have written your summary, reread it with the following question in mind: 'Could I read this precis aloud to the author in the honest belief that it accurately summarises his or her work?' If the answer is 'no', modify your work. You might also ask yourself: 'Would someone who has not read the original text have a good sense of what it is about after reading this precis?' Again, if the answer is 'no', you have some revisions to make.

What is the reader of a summary or precis looking for?

The criteria in the precis assessment schedule in Figure 3.1 can be used as a guide for successfully completing your precis.

Provide full bibliographic details of the text

At the start of your precis, you should provide the reader with a full reference to the work you are summarising. This includes details of:

- author's name (or authors' names)
- date of publication
- title
- edition (if appropriate)
- publisher
- place of publication.

For example:

Northey, M., Knight, D.B. & Draper, D. 2009, *Making Sense in Geography and Environmental Studies*, 4th edn, Oxford University Press, Toronto.

Clearly, if you are writing a precis of a website or web page, you would need to provide the equivalent electronic details. Similarly, a journal article requires additional information, such as the journal name, volume number, issue number and the pages the article appears on. See Chapter 10 for detailed information on correct referencing techniques.

Clearly identify the work's subject matter

After you have written the reference, you should provide a short statement informing your readers what the reviewed work is about. Make it clear, for instance, that you are reviewing a book on 'advances in water quality testing for environmental management' or 'recent developments in political geography'.

Clearly identify the purpose of the work

Having stated what the work is about, you need to let the reader of your summary know precisely what the *aim* of the work is. The distinction between a work's subject matter and its purpose is illustrated by the following introductory passage from Ley and Bourne's (1993, p. 3, emphasis added) edited collection on the social geography of Canadian cities:

> This is a book *about* the places, the people and the practices that together comprise the social geography of Canadian cities. Its *purpose* is both to describe and to interpret something of the increasingly complex social characteristics of these cities and the diversity of living environments and lived experiences that they provide.

In his **preface** to *Recent America*, Grantham (1987, p. x) also distinguishes between the aim and the purpose of the text:

> *Recent America* seeks to provide a relatively brief but comprehensive survey of the American experience since 1945. The emphasis is on national politics and national affairs, including international issues and diplomacy, but some attention is given to economic, social, and cultural trends. I hope that this volume will serve as a useful introduction to a fascinating historical epoch …

You must think carefully about the distinction between a text's subject matter and its purpose. Usually a book or article will discuss some topic or example in order to make or illustrate a particular point or to investigate a specific theme. For example, a volume exploring the transmission of inherited housing wealth to women in Hobart may actually be attempting to contribute to broader discussions of Australia's political economy. You need to be sure you have not confused the subject matter and the purpose.

Be sure to distinguish between the subject matter and purpose of the text you are summarising.

Ensure emphases in the precis match the emphases in the original work

In the summary you should give the same relative emphasis to each area as do the authors of the original work (Northey, Knight & Draper 2009). If, for example, two-thirds of a paper on monitoring water pollution from New South Wales' ski fields is devoted to a discussion of the legalities of obtaining the water samples, your precis should devote two-thirds of its attention to that issue. This helps to provide your reader with an accurate view of the original work.

Ensure that the order of presentation in the precis matches that of the original work

Just as you should give the same relative emphasis to each section, you should also follow the article or book's order of presentation and its chain of argument (Northey, Knight & Draper 2009, p. 67). Make sure you have presented enough material for a reader to be able to follow the logic of each important argument. You will not be able to provide every detail. Present only the critical connections.

If you are reviewing a website, this advice on following the order of presentation may be redundant, given the non-linear structure of some online resources. Instead, consider trying to follow the order of the site as indicated by the layout of links on the site's **home page**.

Summarising a website presents unique challenges.

Outline fully the key evidence supporting the original author's claims

You should briefly mention the critical evidence provided by the author to support her or his arguments (Northey, Knight & Draper 2009, p. 67). There is no need to recount all the data or evidence offered by the author. Instead, refer to the material that was the most compelling and convincing.

Ensure that the precis is written in your own words

Always write the precis in your own words, although you may, of course, elucidate some points with quotations from the original source. Do not construct a precis from a collection of direct quotations from the work you are summarising.

Figure 3.1   Precis assessment schedule

Student Name: Grade: Assessed by:

The following is an itemised rating scale of various aspects of a precis. Sections left blank are not relevant to the attached assignment. Some aspects are more important than others, so there is no formula connecting the scatter of ticks with the final grade for the assignment. Ticks in either of the two boxes left of centre mean that the statement is true to a greater (outer left) or lesser (inner left) extent. The same principle is applied to the right-hand boxes. If you have any questions about the individual scales, comment, final grade or other aspects of this assignment, please see the assessor indicated above.

Description

Full bibliographic details of the text provided				Insufficient bibliographic details
Text's subject matter identified clearly				Text's subject matter poorly or inadequately defined
Purpose of the text identified clearly				Text's purpose not stated or unclear
Emphases in the precis match emphases in original text				Little or no correspondence with text's emphases
Order of presentation in precis matches that of original text				Little or no correspondence with text's order of presentation
Key evidence supporting the original author's claims outlined fully				Little or no reference to original text's evidence
Precis written in own words				Precis constructed largely from quotes

Written expression and presentation

Writing is fluent				Clumsily written, verbose or repetitive
Grammatical sentences				Many ungrammatical sentences

Correct punctuation					Poor punctuation
Correct spelling throughout					Poor spelling
Legible, well set-out work					Untidy and difficult to read
Reasonable length					Too short or long
Correct and consistent in-text referencing style					Incorrect and/or inconsistent in-text referencing style
Reference list correctly presented					Errors and inconsistencies in reference list

Assessor's comments

Writing a good review

A review is an honest, concise and thoughtful description, analysis and evaluation of a work, such as a book, journal article, research report or website. Reviews serve an important role in the professional and academic world, and for that reason you will see book review sections in almost every major journal of environmental studies or geography (for example, *Environment, Geographical Research* and *New Zealand Geographer*). Reviews let people know of the existence of a particular work as well as pointing out its significance. They also warn prospective users about errors and deficiencies (Calef 1964). There are many new publications and online resources appearing, so we need to be selective about what we read.

Reviews are vital, critical guides to new resources.

Lecturers usually ask you to write reviews for one or several of the following reasons:

- to familiarise you with a significant piece of work in the field
- to allow you to evaluate the importance of a work to the discipline you are studying
- to allow you to practise your capacity for critical thought.

What are your review markers looking for?

People who read reviews, including those marking your review, typically want *honest* and *fair* comments on:

- what the reviewed item is about (*description*)
- details of its strengths and weaknesses (*analysis*)
- its contribution to the discipline (*evaluation*).

Figure 3.2 shows the criteria for assessment of a book, article or website review. The next few pages of guidelines and advice will help you to deal with these issues.

A review should be interesting as well as informative. So, while your assessor will probably expect you to deal with the following issues in the course of your review, there are no rules about the order in which you should present them. Instead, you should set out the material in a manner that is both comprehensive and interesting. Before you write your own review, read a few in professional journals in your discipline area to see how they have been laid out and ordered.

Description: what is the reviewed item about?

Description or summary is an important part of a review. You should imagine that your audience has not read the work you are discussing and that their only knowledge of it will come from your review. Give them a comprehensive but concise outline of the work's content and character. However, do not make the mistake of devoting almost all of the review to description—if you do that, your reader might as well go to the original work! As a rule of thumb, try to keep the summary to less than half the total length of your review.

Provide full bibliographic details of the work

You should provide a full and correctly set out reference to the work under review, so that others may consult or purchase it (and so that your marker knows that you have reviewed the correct work). Readers of your review will be interested to know:

- who the work's authors or editors are
- the name of the book's publisher (if it is a book you are reviewing)
- where and when the volume was published.

It is sometimes helpful to state how many pages are in the work. If you are reviewing a book, it is also useful to state its purchase cost, its International

Standard Book Number (ISBN) and whether the volume includes figures and other useful material, although these are rarely required for classroom reviews. Note that a review reference typically contains much more detail than is required in a list of references or bibliography, which generally would not include details such as price or page length (for more details on referencing, see Chapter 10). An example of a complete book review reference is provided below:

> Moulaert, F., Rodriguez, A. & Swyngedouw, E. (eds) 2002, *The Globalized City: Economic Restructuring and Social Polarization in European Cities*, Oxford University Press, New York, xxii + 279pp., plates, tables, figures and index. US$85 cloth (ISBN 0-199-26040-0).

In the above example, the text has 279 pages plus an additional 22 pages of introductory material (for example, acknowledgments, title page, contents pages and notes on contributors). The hardback edition costs US$85. If a paperback existed, its price also would be recorded here. This book also includes photos, tables, diagrams and an **index**. It does not appear to have a bibliography, maps or a glossary. These might have been listed if present.

If you are reviewing a website, you should follow the same bibliographic principles, although the detail will be a little different. An example of a website reference for a review is set out below:

> ABC-CLIO 2000, *World Geography* (online), Available: <www.abc-clio.com>, Date reviewed: 15 November 2000. Market@abc-clio.com, US$499 per year.

Both the author and publisher of this site are ABC-CLIO. Among the details included are the site's **URL** (Uniform Resource Locator or, more simply, its internet address) and the date of the review. The date is useful given the ease and frequency with which websites can be revised. Subscriptions to the site can be ordered through the **email** address provided. This one costs US$499 each year.

Bibliographic details are normally placed at the top of the review. The book review section of almost any geography or environmental management journal will provide an example for you to follow.

Give sufficient details of the authors' background

If you consider it appropriate, and know of the authors' expertise in the area they are writing about, provide a brief overview of their backgrounds and reputations (Marius & Page 2002, p. 213; Northey, Knight & Draper 2009, p. 69). It may also be helpful to consult the web or a *Who's Who* publication to find out a little more about any specific author's affiliation and credentials. You might consider whether the authors have written many other books and articles in this area and whether they have any practical experience. Readers unfamiliar with the subject area will often appreciate some information on the authors' apparent credibility or areas of expertise.

Clearly identify both the work's subject matter and the purpose of the work

See the previous section, 'Writing a Summary or Precis', for details of these requirements.

Correctly describe the author's conceptual framework

Texts are written from a particular perspective. Authors have a way of viewing the world and of arranging their observations into some specific and supposedly comprehensible whole. This way of thinking about the world is known as a **conceptual framework**. You might imagine a conceptual framework to be rather like the work's skeleton, which supports the flesh of words and evidence. As a reviewer, one of your tasks is to expose that skeleton, letting your reader know how the authors have interpreted the issues they discuss. How has the author made sense of that part of the world he or she is discussing?

Try to uncover the intellectual scaffolding that gives the book you are reviewing coherence.

In your review, you might combine identification of the conceptual framework with a critique of it. Does the author make unwarranted assumptions? Are there inconsistencies, flaws or weaknesses in the intellectual skeleton? For example, an author might argue that massive job losses, associated with the adoption of new labour-saving technologies in industry, have been a particular feature of capitalism in Australia. A reviewer, however, might suggest that this reflects a simplistic view of human–technology relations under capitalism and go on to argue that service-sector employment has risen at the same time as industrial job losses have occurred. Moreover, the reviewer might suggest that while some industries have atrophied, others have emerged and grown.

Provide a succinct summary of the work's content

Readers want some idea of what is in the book, article or website. Describe the content in sufficient detail for them to understand what the work is about. This might require stating what is in the various parts of the work and how much space is devoted to each section. You might want to integrate the summary of content with your evaluation of the work, or you might prefer to keep it separate (Northey, Knight & Draper 2009, p. 69). If you are new to review writing, it is usually safer to keep the two separate. Write the summary first, then the analysis.

Accurately identify the intended readers of the work under review

The reader of a review is usually interested to know what sort of audience the author of the original work was addressing. In many cases the author will include a statement on the intended readership early in the volume—often in the preface. For example, in the preface to *The Slow Plague*, Gould says that his book is:

> ... one of a series labeled *liber geographicus pro bono publico*—a geographical book for the public good, which sounds just a bit pretentious until we translate it more loosely as 'a book for the busy but still curious public.'
>
> Gould (1993, pp. xiii–xiv)

Howitt, Connell and Hirsch (1996, p. v) suggest that readers of their edited collection, *Resources, Nations and Indigenous Peoples*, are likely to include students, Aboriginal organisations and mining company staff. In the preface to their book, *Introducing Human Geography*, Waitt et al. (2000, p. vii) explicitly recognise their readers: 'This textbook introduces first-year university students to contemporary themes and practices in human geography.'

Be sure to identify a text's intended audience and review it with that audience in mind.

Figure 3.2   Book, article or website review assessment schedule

Student Name:	Grade:	Assessed by:

The following is an itemised rating scale of various aspects of a review. Sections left blank are not relevant to the attached assignment. Some aspects are more important than others, so there is no formula connecting the scatter of ticks with the final grade for the assignment. Ticks in either of the two boxes left of centre mean that the statement is true to a greater (outer left) or lesser (inner left) extent. The same principle is applied to the right-hand boxes. If you have any questions about the individual scales, comment, final grade or other aspects of this assignment, please see the assessor indicated above.

Description

Full bibliographic details of the work provided					Insufficient bibliographic details
Sufficient details of author's background					No details of author's background
Work's subject matter identified clearly					Work's subject matter poorly or inadequately identified
Purpose of the work identified clearly					Work's purpose not stated or unclear

Figure 3.2   *(cont.)*

Author's conceptual framework identified correctly					Little or no attempt to identify conceptual framework
Work's content summarised succinctly					Excessive/inadequate summary of content provided
Intended readers identified accurately					Readership not identified

Analysis

Work's contribution to understanding of the world/discipline identified clearly					Little or no reference to work's contribution
Clear statement on achievement of work's aims					Work's aims not identified or identified incorrectly
Work's academic/ professional functions identified clearly					Work's functions not identified or identified incorrectly
Work's organisation commented on fairly					Little or no fair comment on organisation
Work's evidence evaluated critically					Little or no critical evaluation of evidence
Work's references evaluated critically					Little or no critical evaluation of references
Style and tone of presentation evaluated critically					Little or no critical evaluation of style and tone
Quality of supplementary material (e.g. tables, maps, plates) reviewed competently					Little or no critical evaluation of supplementary material
Other deficiencies/strengths in the work identified correctly and fairly					Other evident weaknesses/strengths not identified

Evaluation

Work compared usefully with others in the field					Little or no effort to compare work with other works in the field
Valid recommendations made on the value of reading the work					No recommendation provided or recommendation inconsistent with earlier comments

Written expression, references, and presentation of the review

Various sections of the review of appropriate length					Major imbalances evident
Fluent piece of writing					Clumsily written, verbose or repetitive
Grammatical sentences					Many ungrammatical sentences
Correct punctuation					Poor punctuation
Correct spelling throughout					Poor spelling
Legible, well set-out work					Untidy and difficult to read
Reasonable length					Too short or long
Correct and consistent in-text referencing style					Incorrect and/or inconsistent referencing style
Reference list correctly presented					Errors and inconsistencies in reference list

Assessor's comments

Identifying the intended audience of the reviewed work serves at least two important purposes. First, you will be helping your readers to decide whether the original work is likely to be of any relevance to them. Second, you will be providing yourself with an important foundation for writing your critique of the work. For example, from Gould's statement above, it is reasonable to conclude that his book ought to be easy to read, stimulating and written for a lay audience. If it is not, there is an important flaw in the book. A book or article intended for experts in the field may legitimately use technical terms and express complex ideas. A version of the same information for children, on the other hand, may be simplified with a completely different vocabulary. You should write your review keeping in mind the relationship between the intended audience and the book's content and style.

Analysis: details of strengths and weaknesses

So far you have let your reader–assessor know a few basic descriptive details about the work you are reviewing. Now you need to let everyone know what you consider to be the weaknesses *and* strengths of the work. Being negative about a volume under review does not necessarily suggest that you are smarter than the author. Indeed, it can be more difficult and challenging to demonstrate how good a work is. If you believe that the material you are reviewing has no significant weaknesses, you should say so. However, you should also point out its specific strengths. As Calef (1964, p. 1) observes, analysis is usually the weakest feature of book reviews:

> All authors deserve sympathetic, appreciative analyses of their books; too few authors get them. Many reviewers concentrate on the authors' mistakes and discuss the books as they should have been written.

When you write your review, it is essential that you consider the aims of the authors. *Analyse the work on the authors' terms.* Have the authors achieved their aims?

Be fair, be explicit and be honest when you write a review.

In your analysis you should, above all, 'be fair, be explicit, be honest' (Calef 1964, emphasis added). To these ends you should explain why you agree or disagree with the authors' methods, analysis or conclusions.

In organising their analysis of a work, many reviewers first point out the utility and successes of the book, article or website and then move on to point out its deficiencies. You may find that pattern a useful one to follow.

Clearly identify the work's contribution to your understanding of the discipline

In evaluating the work's contribution to the discipline, you should begin with the assumption that the author has something useful to say, rather than trying to explain whether you agree or disagree with that contribution. What is that contribution? Has the author helped you to make sense of things? What has been illuminated?

State clearly whether or not the work achieved its aims

Think carefully about the author's objectives and compare them with the content of the work. Do they match one another? You would be derelict in your duty as a reviewer if you had stated what the reviewed work's aims were, but failed to say whether or not they had been met. (Imagine how frustrated you would feel if someone told you there was an article in a magazine about how to make a lifetime fortune in twenty-one days, and when you read those pages you found that the article failed to deliver what it promised.)

Clearly identify the work's academic or professional functions

Ask what educational, research or professional functions the reviewed work might fulfil. For example, is the book, article or website likely to be a useful resource for people in the same class as you, for other undergraduate students, or for leaders in the field?

Accentuate the positive. For example, an author may think that the book's audience ought to be final-year undergraduate students, but you—as an undergraduate student reviewer—believe the book would better serve a first-year audience. Rather than simply stating that the work is inappropriate for its intended audience, let your reader know which audience the work might best serve.

Try to be helpful, rather than negative, in your analysis.

Comment fairly on the work's organisation

Say whether you believe the reviewed work is well organised or not. In this context, think about the ways in which the book or paper is subdivided into chapters or sections, or how easy it is to navigate around the website. Do the subdivisions, chapters or links advance the work's purpose or are they obstructive? Do they break up or upset an intellectual trajectory? Support your viewpoint with examples.

Critically evaluate the work's evidence

An author's evidence should be reliable, up to date and drawn from reputable sources (such as official statistics, *Hansard*, international journals, experts in the field or links to credible and relevant websites). It also should support claims made in the work. Would the results of the original work stand up to replication? That is, if the study were done again, is it likely that the results would be the same? Would the results stand up to corroboration? That is, are the results of the study substantiated by other related evidence?

Give a clear assessment of the evidence used in the work. Is the evidence compelling, or not? If, for example, you are sceptical about the repeated use of quotations drawn from *National Enquirer* to support a discussion of gerrymandering and electoral bribery in Queensland, you should let the reader know. Back up your assessment of the evidence with reasons for your conclusion. If you are able, and if it is necessary, suggest alternative and better sources of evidence.

Critically evaluate the work's references

Readers and assessors of your review will wish to know if the reviewed work has covered the available and relevant literature to a satisfactory degree. If you are reviewing a website, you might expand your notion of references cited to include hotlinks from the reviewed site to other relevant websites and pages. If there are major shortcomings in the references acknowledged, it is possible that the work's authors may not be fully aware of material that might have illuminated their work. If you are new to your discipline you might protest, sometimes quite justifiably, that you cannot offer a meaningful judgment about the strength of the reference material. Nevertheless, you ought to be thinking about this question and answering it where possible.

Critically evaluate the style and tone of presentation

Among other things, readers of a review may be trying to work out whether to buy, read or consult the reviewed work. Therefore, questions foremost in the mind of some will be whether the work is written clearly and if it is interesting to read. Is the writing repetitious or boring? Is it detailed or not detailed enough? Is the style clear? Is it tedious, full of jargon or offhand? Particularly if you think there are problems with the style of writing, it is appropriate to support your criticisms with a few examples.

If you are reviewing a website, you might also consider commenting on the visual character and quality of the site. For example, is judicious use made of technical features or has wizardry stolen the show?

You should also comment on the tone of the work. Let your reader or assessor know if the work is only accessible to experts in the field or if it more resembles a 'coffee table' publication. Of course, criticisms about the tone of the work should be written with reference to the intended audience. It would be unfair, for instance, to condemn a text on the grounds that it used technical language if it was written for an audience of experts.

Ensure that the quality of supplementary material is reviewed competently

Many books, articles and websites make extensive use of supplementary material such as tables, plates, maps and figures. In your review, you should comment on the quality of these materials and their contribution to the work's message. The graphic and tabular material should be relevant, concise, large enough to read and comprehensible, and the details of sources should be provided.

Correctly and fairly identify any other deficiencies and strengths in the work

After a close reading of the work, you should be able to identify any weaknesses and strengths you have not already discussed. Remember: you do not *have* to find things wrong with the work you are reviewing. Being critical does not require you to be negative if it is not warranted. Similarly, you should not pick out minor problems within the work and suggest that they destroy it entirely. However, if there are genuine problems, be explicit about what they are, providing examples where possible.

Support criticisms in your review with evidence.

Evaluation: contribution to the discipline

In the first section of the review you described the work being reviewed. You then went on to outline its strengths and flaws. Now you have to make a judgment. Is the work any good? You may find it useful to be guided by a central question: 'Would you advise people to read (or buy) the work you have reviewed?'

Compare the work's usefulness with that of others in the field

If you have sufficient expertise in the field, **appraise** the work being reviewed in terms of its use as an alternative to work already available. It may be helpful to consult library reference material to see, for example, how many other volumes on the same (or a similar) topic have been produced recently. If you are reviewing a website, search for competitors and compare them.

Think about how the text you are reviewing stacks up against its competition.

Compare the work you are reviewing factually with its predecessors. What subjects does it cover that earlier volumes did not? What does it leave out? (Northey, Knight & Draper 2009, p. 69) Remember to correctly cite any additional sources you use.

Make a valid recommendation on the value of reading the work

A fundamental reason for writing reviews is to let readers know whether a particular book, article or website is worth consulting. Your recommendation should be consistent with the preceding analysis of its strengths and weaknesses. For example, it would be inappropriate to **criticise** a work mercilessly and then conclude by saying that it is an important contribution to the discipline and should be consulted by everyone interested in the area. It is worth restating a point made earlier: in your review 'be fair, be explicit, be honest'.

Written expression, references and presentation of the review

Make sure the various sections of the review are of appropriate length

Assessors will consider the balance between description, analysis and evaluation when marking a review. A description of the work is important, but it should not dominate your review. As noted earlier, unless there are special reasons, the description usually should be less than half of the total length of the review.

Your own writing should be clear, concise and appropriate for your audience. Detailed advice on the assessment criteria used for written expression (shown in the assessment schedule for reviews in Figure 3.2) can be found in Chapter 1.

Some examples of reviews

Before you start writing a review, you may find it helpful to look to high-quality academic journals (for example, *Area*, *Geographical Research* or *Professional Geographer*) for good examples. In Box 3.3, and as a preliminary guide, are some slightly revised versions of reviews published in the *library journal*. You will see that, despite their brevity, the reviews address many of the issues discussed in this chapter. The last example is a review of a website.

Box 3.3 Examples of reviews

Example 1

Mommsen, H. 1996, *The Rise and Fall of Weimar Democracy*, trans. E. Forster & L. E. Jones, University of North Carolina, Chapel Hill, pp. 608.

This translation makes available to the English-speaking world an important historical work published in 1989 by a prominent German historian. The period from 1919 to 1933 was a time of great political, social, economic and artistic upheaval in Germany. In this magisterial work, the author looks at the Weimar period from the viewpoint of social and economic history, providing a lucid yet detailed account of the complexities of this era without attempting to pull the strands out of their context in order to find the 'roots' of the Nazi period that followed. By doing so, he both sheds light on this interesting period and provides a convincing overall picture of Germany's road from parliamentary democracy to dictatorship during the interwar years. Because of its depth and detail, this book is not for casual readers, but it covers a complex subject with admirable clarity and is certain to become a classic. Everyone interested in European history should read this book.

Adapted from: *library journal* (1996, vol. 121, no. 12) © library journal, 1996.

Example 2

Meredith, M. 2005, *The Fate of Africa: A History of Fifty Years of Independence*, PublicAffairs Perseus Publishing, New York, pp. 800.

A scholar of Africa necessarily becomes an expert on death. In Meredith's tome, death comes in huge numbers and in many ways: through famine, ethnic strife and racial injustice, and at the hands of ruthless dictators. It came in the days of European colonialism, but in postcolonial Africa, death pervades the continent. Meredith (*Our Votes, Our Guns: Robert Mugabe and the Tragedy of Zimbabwe*) writes with sobriety, intelligence and a deep knowledge of Africa as he describes individuals responsible for deaths unimaginable to much of the rest of the world. A well-known example is the carnage among Hutus and Tutsis in Rwanda, which claimed 800,000 lives in 100 days in 1994—more people were killed more quickly than in any other mass killing in recorded history. Much of this tragic history has been told in part elsewhere, but Meredith has compiled a text covering the entire continent. Only in the last few pages does Meredith answer the question of Africa's fate—and he thinks it's bleak. This is a valuable work for those who wish to understand Africa and its besieged peoples.

Adapted from: *library journal* (2005, vol. 130, no. 12) © library journal, 2005.

Example 3

Rossi, J. 2000, *The Wild Shores of Patagonia: The Valdes Peninsula and Punta Tombo*, Abrams, New York, pp. 224.

Neither an experienced author nor a photographer, Rossi discovered Patagonia while on short leave from her job with the European Parliament and promptly turned her fascination into a two-year encampment to prepare this book. About half the book is devoted to a detailed discussion of southern right whales, southern elephant seals and killer whales (including their unique beach hunting methods). Less detail is given to South American sea lions, and dusky and bottlenose dolphins. The author describes Magellanic penguins, but provides limited information on 12 other kinds of birds, sometimes furnishing just one picture and caption per species. Rossi also discusses eight land animals in about 25 pages, giving short shrift to some unique creatures. Reader interest in this area may grow, since the United Nations has recently declared the Valdes Peninsula a World Heritage Site, and ecotourist visits are increasing. Although the photographs are merely average, this readable volume is recommended for members of the general public and for academic audiences interested in environmental and life sciences.

Adapted from: *library journal* (2001, vol. 126, no. 1) © library journal, 2001.

Example 4

ABC-CLIO 2000, *World Geography*, ABC-CLIO, Available: <www.abc-clio.com>, Date reviewed: 15 November 2000. Market@abc-clio.com, US$499 per year.

World Geography aims to be the place students go to research a number of themes and look up facts (many nongeographical) on all countries. There are three major sections providing access to the data: Home, Student and Reference. The site includes more than 10,000 entries, including biographies, histories, maps, documents, statistics, video clips and photographs. In the Home section, students can select a country to study; search the text by keyword or choose to catch up on world events; read a feature article ('Why is the Price of Gas Rising?'); or answer three questions to a 'Where in the World?' quiz. The feature article portion is well researched; one hopes ABC-CLIO plans to archive these essays.

In the Reference section, country selection is via a world map or drop-down menu. Text can be searched by keyword or in advanced search, where searches can be limited by type. A noteworthy inclusion here is 'Ask the Cybrarian', in which students can ask reference questions of ABC-CLIO.

The Student section can be used in the classroom. Students can read the syllabus, class announcements and the word of the day, and take review tests. A nice feature here is regional overviews that cover Landforms, Climate, Vegetation and Animal Life, People, and Natural Resources and Agriculture. Other overviews cover people, news events and organisations for the selected country.

There are also a few problems. Though country maps are clear, city maps are unfortunately excluded. Data presented are only for one year, although often more historical data are needed. News events in the Home section are listed incompletely; for example, 'Experts Investigate Possible ...' Possible what? Forcing users to click to see if a story is worth reading is inefficient. The icon used to designate a link to a map is a small map of the United States—for all countries. Sources for news articles are not always provided, nor are clear directions on how to cite information found here. Information included is sometimes mystifying; for example, under Organizations in the United States section, corporations such as Apple, Walt Disney, AT&T and Exxon are included. Pierre Trudeau's death is noted, but no biography is available. The Tools section offers two excellent features: the Merriam-Webster's Collegiate Dictionary and the Merriam-Webster's Collegiate Thesaurus, and ClioView allows students to select, sort and view a country and topic (with 18 from which to choose, including Literacy, Population Density and Active Armed Forces).

The bottom line: World Geography is not exactly what its name implies. It is ABC-CLIO's successor to Exegy and is better described as a good source for beginning country studies. In that context, it is a useful resource. This site is recommended for members of the general public and for children entering secondary school.

Adapted from: *library journal* (2001, vol. 126, no. 1) © library journal, 2001.

REFERENCES AND FURTHER READING

Berry, B.J.L. 1993, 'Canons of reviewing revisited', *Urban Geography*, vol. 15, no. 1, pp. 1–3.

Brozio-Andrews, A. 2007, *Writing Bad-book Reviews*, viewed 12 May 2011, <www.absolutewrite.com/specialty_writing/bad_book_reviews.htm>.
A very brief resource that differentiates between a bad book review and a review of a bad book.

Burdess, N. 1998, *The Handbook of Student Skills for the Social Sciences and Humanities*, 2nd edn, Prentice Hall, New York.
Chapter 3 includes short sections on writing abstracts and book reviews.

Calef, W.C. 1964, *Canons of Reviewing*, Illinois State University.

Callicott, J.B. & Frodeman, R. (eds) 2008, *Encyclopedia of Environmental Ethics and Philosophy*, vol. 2, pp. 507–14, Gale Cengage Learning, Detroit.

Clanchy, J. & Ballard, B. 1997, *Essay Writing for Students: A Practical Guide*, 3rd edn, Addison Wesley Longman, Melbourne.
Includes a succinct description of how to write a review.

Colorado State University 2011, *Writing Guide: Book Reviews*, viewed 12 May 2011, <http://writing.colostate.edu/guides/documents/bookreview/>.
A very (perhaps overly) detailed outline of book review writing.

Cottingham, C., Healey, M. & Gravestock, P. 2002, *Fieldwork in the Geography, Earth and Environmental Sciences Higher Education Curriculum*, viewed 12 May 2011, <www2.glos.ac.uk/gdn/disabil/fieldwk.htm>.

Engle, M., Blumenthal, A. & Cosgrave, T. 2011, *How to Prepare an Annotated Bibliography*, Cornell University, viewed 12 May 2011, <www.library.cornell.edu/olinuris/ref/research/skill28.htm>.

Federal Railroad Administration, Office of Policy 1992, *Transportation and the Environment: An Annotated Bibliography*, viewed 12 May 2011, <http://ntl.bts.gov/DOCS/tea.html>.

Galvan, J.L. 2009, *Writing Literature Reviews: A Guide for Students of the Social and Behavioral Sciences*, 4th edn, Pyrczak Publishing, Glendale, California.
An easy-to-follow guide offering advice and examples.

Gould, P. 1993, *The Slow Plague: A Geography of the AIDS Pandemic*, Blackwell, Cambridge, Massachusetts.

Grantham, D.W. 1987, *Recent America*, Harlan Davidson, Arlington Heights.

Hadden, R.L & Mincon, S.G. 2010, *The Geology of Haiti: An Annotated Bibliography of Haiti's Geology, Geography and Earth Science*, US Army Corps of Engineers, Alexandria, Virginia.

Henderson, E. & Moran, K. 2010, *The Empowered Writer. An Essential Guide to Writing, Reading, and Research*, Oxford University Press, Don Mills, Ontario.
A very thorough guide to high-quality academic writing, with detailed advice on summaries set out in Chapter 9.

Howitt, R., Connell, J. & Hirsch, P. (eds) 1996, *Resources, Nations and Indigenous Peoples: Case Studies from Australasia, Melanesia and Southeast Asia*, Oxford University Press, Melbourne.

Kamerman, S.E. (ed.) 1978, *Book Reviewing*, Writer, Boston.
Now dated, this large volume on book reviewing written by experts still offers helpful advice. See especially the chapters 'Do's and don'ts of book reviewing', 'The basics of book reviewing' and 'The importance of book reviewing'.

Ley, D.F. & Bourne, L.S. 1993, 'Introduction: The Social Context and Diversity of Urban Canada', in *The Changing Social Geography of Canadian Cities*, eds L.S. Bourne & D.F. Ley, McGill-Queen's University Press, Montreal & Kingston.

Marius, R. & Page, M.E. 2002, *A Short Guide to Writing About History*, Longman, New York.

Northey, M., Knight, D.B. & Draper, D. 2009, *Making Sense in Geography and Environmental Studies*, 4th edn, Oxford University Press, Toronto.
See Chapter 6, especially, which provides advice on writing summaries (precis), analytic book reports and literary reviews (i.e. focusing on a theme and requiring coverage of several books).

Ormondroyd, J., Engle, M. & Cosgrave, T. 2011, *Critically Analyzing Information Sources*, viewed 12 May 2011, <http://olinuris.library.cornell.edu/ref/research/skill26.htm>. *Another useful resource from Cornell University, this distinguishes helpfully between an initial review of a source's bibliographic details and a more comprehensive review of its content.*

Reed, T.V. 2011, *Environmental Justice Cultural Studies Bibliography*, viewed 30 May 2011, <http://culturalpolitics.net/environmental_justice>.

Waitt, G., McGuirk, P., Dunn, K., Hartig, K. & Burnley, I. 2000, *Introducing Human Geography*, Longman, French's Forest.

Weintraub, I. 1994, 'Fighting environmental racism: a selected annotated bibliography', *Electronic Green Journal*, vol. 1, no. 1, unpaged.

Writing Center, University of North Carolina at Chapel Hill 2007, *Book Reviews*, viewed 12 May 2011, <www.unc.edu/depts/wcweb/handouts/review.html>. *A very helpful summary resource discussing process and strategies for writing good book reviews.*

Zeegers, P., Deller-Evans, K., Egege, S. & Klinger, C. 2008, *Essential Skills for Science and Technology*, Oxford University Press, South Melbourne. *Chapter 15 provides some helpful and succinct advice on writing an article review, and includes a template for reviewing empirical articles.*

FOUR

Writing a media release

How is the world ruled and how do wars start? Diplomats tell lies to journalists and then believe what they read.

Karl Kraus

Key topics

- What is a media release and what are users of media releases looking for?
- Targeting a media release
- Writing a media release
- Sending a media release
- Following up a media release

Geographers and environmental managers are often interested in ensuring that the media, and thereby the general public, know of their work and their views on important issues. This chapter provides some guidance on how to prepare material for the mass media, and offers some short comments on conducting a successful radio interview. Although internet-based communication with broad audiences is emerging as an increasingly important means of communicating geographical and environmental messages, this chapter does not set out advice on the diverse modes of such self-publishing. These expanding varieties of communication might take the form of building your own website, blogging, contributing to user-generated social news websites (such as NowPublic) and using social and professional networking sites (such as Academia.edu, Bebo, Facebook and LinkedIn). Specific advice on using these resources is available online, often through the websites themselves.

Despite the burgeoning presence of these online social media and professional networking sites, the traditional media—press, radio and television—remain powerful tools to access the general public. However, before you decide which media you wish to target, it is important to identify exactly why you want to access the media in the first place. Box 4.1 offers a selection of reasons.

The mass media provide access to vast, valuable and powerful audiences.

> **Box 4.1 Why write for the media?**
>
> There is a wide variety of reasons for writing for the media. A written piece may:
>
> - inform and raise awareness among the general public about geographical and environmental issues
> - communicate the results of your scholarly activities to broad and interested audiences
> - offer your selection of facts and views on a particular situation or issue (that is, agenda setting)
> - help to establish a favourable public profile for your work, your discipline and/or the organisation for which you work; this may help to find sponsors and other forms of support
> - promote and secure support for activities (for example, presentations, public meetings, research in progress and funding)
> - help to develop your local profile as someone with something useful to say about a specific topic—to become an 'authorised knower' (Hay & Israel 2001)
> - help reporters to get their facts right.

What is a media release and what are users of media releases looking for?

The most commonly employed means of communicating with the media is through a **media release** (also known as a press release). Media releases are a major source of information used by journalists. They generally set out the core information about something you hope will get favourable editorial coverage.

A media release is often your first contact with the media about any story. It also may be your only contact. You should try to ensure that the release is well written, comprehensive and captivating. Try to provoke interest and set the tone for the ways in which the story might subsequently be covered. You need to ensure that your release in some way meets the needs of media professionals, as discussed in Box 4.2.

**Box 4.2
What are users of media releases looking for?**

Remember, mass media is essentially a profit-driven business. Staff at newspapers, radio stations and television stations aim to sell as much of their news product (and associated advertising in the case of commercial outlets) to as many readers, listeners or viewers as possible. For this reason, they are most likely to be interested in stories that are:

- accurate
- new or up to date (for example, an economic geographer's explanation for a recent rise or fall in national currency)
- unusual or unexpected (for example, the discovery of a new species of tuatara on an offshore island of New Zealand, or the case of a heterosexual couple who are thrown out of a gay bar for kissing publicly)
- captivating (for example, an oceanographer's professional views on how to make America's Cup yachts sail faster by taking advantage of local knowledge)
- relevant to a large part of the medium's audience (for example, a forecast by an urban geographer of plummeting real estate values in Sydney, which is likely to be of deep interest to readers of the *Sydney Morning Herald* and perhaps the Melbourne *Age*, but for quite different reasons!)
- relevant to a local or regional audience (most media outlets have specific markets to which they aim to make their stories interesting and relevant. Thus, a small country town's newspaper may carry stories on environmental issues relating to the surrounding region that a larger metropolitan paper will not be interested in covering).

Targeting a media release

A media release is a vital bridge between your academic and professional work and the mass media.

Despite all of the above, the success of a news release also depends on it being directed at the correct outlet. You are unlikely to get much coverage if you direct your great story about revolutionary applications of Global Positioning Systems (GPS) for sheep tracking in the New Zealand high country to the Brisbane *Courier Mail*! Accordingly, you should consider familiarising yourself with media outlets you think might be relevant to your work and interests. Assess the market. Box 4.3 lists some questions to consider when deciding how to target your media release.

Direct your attention to matters such as:

- What relevant papers and magazines exist?
- How often do these publications appear (daily, weekly, monthly, quarterly or annually)?
- To what audiences is the content relevant (for example, local, state, national or international; conservative, wealthy or left-wing)?
- How much space or time is devoted to specific issues?
- Do the stories that are run challenge or uphold the status quo?
- What are the characteristic styles of writing used by the outlet?

**Box 4.3
Targeting
a media
release**

Check the regular features pages or sections within the publications you are targeting. Many metropolitan newspapers have special sections on specific days of the week (for example, sport on Monday, environment on Tuesday, real estate on Wednesday, higher education on Thursday and so on). Not only might this be of significance in terms of the timing of your press release, but it will also point out the sorts of issues the publication considers most important.

Selecting the right media outlet is vital to the success of your story.

Writing a media release

There are several conventions to follow when creating a media release. These allow media workers under deadline pressure to quickly assess a release's newsworthiness and adapt it to their needs.

Layout and presentation

- Type the media release on one side of a sheet of paper, double-spaced, justified left and right with wide margins. This allows it to be subedited and passed on immediately to a printer or announcer.
- If you are representing an organisation or institution, use their letterhead paper. If this is not available to you, be sure to type your name, that of the organisation with which you are associated, and full address and contact details at the top.
- Date the media release.
- Mark the document 'media release' and add a short, attractive title that explains what the release is about.
- Number the pages of the release and associated information.

Structure

In many respects, a media release does not follow the principles of good writing. It is certainly different from other forms of written communication with which you might become familiar at university. It does, however, require attention to structure. The media release must catch the reader's attention (hook the reader), provide a context and deliver easily digested quotes. In writing a media release it is often tempting to write too much. Before you can access potential readers, you have to get the message past the desks of the journalist and editor. This is not an easy task as even small circulation papers may receive forty to fifty releases per day (Minnis & Pratt 1995).

A media release follows an inverted pyramid structure so that whole paragraphs near the end can be deleted without losing the story's central message.

- Use a catchy headline and set it out in large, bold capital letters (for example, **CAT LOVERS SPITTING MAD**).

- Make your most important points first. Typically, news stories are written according to a fairly rigid top-down formula, known as the 'inverted pyramid'. They begin with the most important, central information and then go on to provide explanatory paragraphs with other details in declining order of importance. There are several reasons for this. First, the reader's attention is captured immediately. Second, because most people do not have time to read the entire newspaper, it is helpful to them if they can gain a broad overview of the news of the day by reading headlines and introductory paragraphs. Third, stories often have to be edited down to fit available space. With the inverted pyramid structure, subeditors can be fairly confident that by removing material from the end of the story, they are not removing any vital information. For this reason, check any release you write to ensure that it will still make sense if paragraphs are removed from the end.

- Keep the opening paragraph (the 'intro') brief. Aim for fewer than 40 words. Williams (1994) suggests the ideal length is 17–25 words. Ensure the intro makes an impact and highlights the newsworthy angle of the story. This will encourage you—and anyone who reads the release—to focus on the main story hook.

- The opening paragraph provides the 'story hook' (Williams 1994) and should include the *who, what, when, where* and *why* of a story. For example, 'Angered by recent proposals to end domestic cat ownership by 2030 [WHY], the Cat Appreciation and Taming Society (CATS) [WHO] has convened a public meeting [WHAT] to discuss new pet ownership regulations. The meeting will be held at the Onsley Town Hall [WHERE] on 12 December [WHEN].' If you deal with the 'so what' issue, so much the better; that is, see if you are making it clear why someone reading the release should give it any further attention.

- The second paragraph can provide a 'secondary hook' (Williams 1994) and should explain in more detail the information from the first paragraph. For example, 'CATS has convened the meeting in response to radical proposals by the Member for Onsley, Mr Jim Nopsis, to make it illegal to own cats in the state.'
- Keep paragraphs short. Try to limit them to a single sentence.
- The release should conclude with 'END' so the journalist can differentiate between the media release and subsidiary information, such as contact details and photographs.
- End the release with the names and full contact details of a minimum of two reliable people who can be contacted for more details. It is helpful to include daytime and evening phone numbers. If the journalist finds no one to talk to about the release, the story may be abandoned. After-hours contact details, including weekends, are essential.

Make yourself available to journalists following up on your release.

Evidence and credibility

- Near the top of your story, provide a few details about yourself or your organisation, such as the function it serves and its number of members.
- Do not assume reporters will be familiar with you or the organisation with which you are associated. Mention its full name early in the release. If it has an acronym, add this in parentheses.
- Use quotes wherever possible. These give the story vitality and credibility. Be aware that many quotes in media releases are in fact entirely manufactured. No one actually said them! If this is the case with your release, at least run the quote past the person to whom the quote is attributed to see if they are happy with its style and content.
- Attribute statements to a particular person and not to an anonymous spokesperson. Give the spokesperson's full title, name and position (for example, Dr Felicity Sealpoint, President of the Cat Appreciation and Taming Society). It is often helpful if the people quoted are prominent within the community in some way or are leaders in their field.
- Provide each person's full name and be sure to include some phrase that makes it clear to readers what that person's claim to authority in this story is (for example, 'Local breeder of Singapura cats, Mrs Mee Yeo, said ...').
- Support statements and claims with 'hard' evidence such as statistics, quotes and references to government documents and other reports.
- Always be accurate.

Quotes from key figures lend authority to a media release.

Language

- Ensure the release is clear, concise and arresting. Try to keep it to a page or less (that is, about 300–400 words). Keep the sentences short. Leave out unnecessary words. Readership can fall off by as much as 80 per cent after the third or fourth paragraph of a news story (Williams 1994, p. 6). With this in mind, short and to-the-point releases should be the goal.
- Keep language simple and concise (for example, use 'now' rather than 'at this point in time'; and 'here' rather than 'hereabouts').
- Use the word 'said' when attributing quotes to people. Although there is a wide variety of options, such as 'exclaimed', 'declared', 'stated', 'expressed' and so on, try to avoid them. Although they may not be aware of it, most newspaper readers expect to see 'said'. Anything else may prove disruptive.
- Avoid jargon and clichés 'like the plague'!
- Use colourful quotes. A statement like, 'The Honourable member for Onsley is a despicable cat-killer' is likely to get good coverage—though you should be quite sure it is true if you plan to use something quite so inflammatory! Do let people know that you plan to quote them.
- Ensure all spelling is correct and that all names and titles are accurate.
- Unless the story is about something that is yet to happen, use the past tense.
- Use the active rather than the passive voice. In the passive voice, the subject receives the action expressed by the verb (for example, the tree was hit by the moving car), whereas in the active voice the subject performs the action of the verb (for example, the moving car hit the tree). In general, the active voice is more lively and succinct.
- Use positive rather than negative statements.

Engaging with the media can be tiresome and disheartening. A media release that may have taken many hours for you to construct may not even make it beyond the wastepaper bin of a newspaper office. If the story does make it into the public arena, it may look nothing like your original. Walters, Walters and Starr (1994) reported that, on average, press releases were halved in size by journalists and editors from a mean of 434.8 words to 209.8; 13.5 paragraphs to 6.3; and 18.4 sentences to 9.3. The reading level of releases was lowered ('dumbed down'), with the Flesch-Kincaid-Grade level reduced from 14.98 to 13.60 and the average length of words reduced from 5.28 to 5.05 letters. In these cases, the journalist and editor are making your story more palatable and hence more digestible for the audience. Do not regard the changes as a slight against your professionalism or integrity. It is simply that the journalist and editor will have a more highly developed sense of the language level most appropriate to their readers.

Photographs and supplementary information

- It can be very helpful to submit a high-quality, relevant photograph (or other illustrative material such as a map) with your story, particularly if you are submitting it to a small media outlet with limited resources. Digital images should be of high quality and submitted in a standard format such as JPG, TIF or PNG. If hard-copy photographs are more appropriate, these also should be of very high standard, and about 150 × 100 millimetres in dimension. While larger newspapers may not use your photo, it might give their staff some ideas for pictures of their own. Include a **caption** for the photograph (ensure the caption answers what, where, when and who questions). If you are using a hard-copy image, attach the caption to the back of the picture.
- In some cases a picture can make a story. For instance, a recent close-up photograph of a Tasmanian tiger (*Thylacinus cynocephalus*), regarded as extinct since 1936, is likely to make the front page of most Australian national newspapers, whereas an unillustrated report of its sighting is unlikely to figure prominently in any newspaper.
- Try to avoid large group photos. When reduced in size for publication, these tend to lose relevant details.
- If you find that the complexity of the story warrants it, consider including with your release a separate 'fact sheet' that explains complicated matters clearly. If your release describes a paper or research report you have written, it may be helpful to send in a copy of the report with your press release. Do not expect to receive the report back, though!

Consider including a high-quality, relevant photo with your media release.

Timing matters

- Give the journalists plenty of time to react to your media release (days rather than hours if possible).
- Daily papers are typically busiest in the morning and early afternoon. For morning papers, the deadline for most stories is late afternoon (only major stories are likely to be published after that). Unless you have a story of tremendous significance and urgency, do not call a journalist after about 4 p.m. You are unlikely to get a warm reception. The best time is likely to be between noon and 2.30 p.m. Contact staff on evening papers between 7.00 a.m and 8.30 a.m.

Timing your media release can be critical to its success, so learn about your local media schedules.

- Weekly papers, such as those published on Sundays, typically face a Wednesday or Thursday deadline (unless the story is of major note).
- Consider sending your news release to newsrooms on Saturday or Sunday as these are usually the quietest days of the week. This will require you to provide after-hours contact details on your release. Mobile telephones can be a real bonus here, but make sure your batteries are charged and you have good coverage!
- On public holidays and in major holiday seasons such as Christmas and early January (in the southern hemisphere), newsrooms are often desperate for new stories.
- Where possible, try 'piggybacking' your story onto some significant local or international event. If, for instance, you have been conducting research work on strategies to curb the impact of feral cat populations on native fauna, and you see the 'Cat Lovers Spitting Mad' article in your local paper, it might be timely to send in a news release about your work. Keep a close eye on those current events that might make your release topical.
- While piggybacking may be helpful, try to avoid clashing with a foreseeable event of such magnitude that it will dominate the news.
- If you have any concerns about when the release should be published, state at the top either 'For immediate release' or if you prefer later public disclosure, say '**Embargoed** until [time], [date]'.

Media release format

Although not all media releases will follow it exactly, the basic outline for a media release is shown in Box 4.4.

Box 4.4 Suggested layout of a media release

Letterhead
FOR IMMEDIATE RELEASE [or Embargoed until (time), (date)]
Date
Headline
Introductory paragraph that answers *Who*, *When*, *Where*, *What* and *Why*?
  Second paragraph explaining in more detail the information from the first paragraph.
  Third paragraph that includes a quote attributed to some prominent person.
  Fourth paragraph that includes some more information, perhaps another quote.
  Subsequent paragraphs that add supplementary information that can be removed by a subeditor without making the earlier parts of the release incomprehensible.
-END-

For Further Information Contact:

Full title and name of contact	Full title and name of contact
Direct phone number (BH)	Direct phone number (BH)
Direct phone number (AH)	Direct phone number (AH)
Mobile phone number	Mobile phone number
Email address	Email address
URL [if appropriate]	URL [if appropriate]

Following are examples of: (a) a fictitious media release written according to this format; and (b) a verbatim copy of a release prepared in an Australian university to bring public attention to work carried out there on the geographical dimensions of racism.

The Cat Appreciation and Taming Society
27 March 2012
MEDIA RELEASE for immediate release
CAT LOVERS SPITTING MAD

Adelaide, Australia. Angered by recent proposals to end domestic cat ownership by 2030, the Cat Appreciation and Taming Society (CATS) has convened a public meeting in Onsley to discuss new pet ownership regulations.

CATS has convened the meeting in response to radical proposals by the Member for Onsley, Mr Jim Nopsis, to make it illegal to own a cat after 2030.

According to CATS, Mr Nopsis's proposals arise from misguided concerns about the effect of cats on native wildlife.

Commenting on the proposed regulations, CATS president Dr Felicity Sealpoint described Mr Nopsis as 'an ill-informed and despicable cat killer'.

'Not only are these proposals frighteningly cruel, they are also a flagrant infringement on our civil rights,' Dr Sealpoint said.

The CATS meeting will be held in the Onsley Town Hall at 7 pm on 1 April.

-END-

For further media enquiries please contact:
Dr Felicity Sealpoint
President, Cat Appreciation and Taming Society
Tel: (08) 8123–9876 (BH), (08) 8765–1234 (AH), 0123 123 123 (Mob)
Email: sealpoint@tinroof.com

Box 4.5
Example of a fictitious media release

Or
Mr Rex Cornish
Secretary, Cat Appreciation and Taming Society
Tel: (08) 8381–3456 (BH), (08) 8123–4567 (AH), 0321 321 321 (Mob)

**Box 4.6
Example
of a media
release
about
geographical
research**

THE GEOGRAPHY OF RACISM

The occurrence of racism in NSW is regionally specific, according to new research from UNSW. The regional variations are associated with average educational attainment, the distribution of cultural groups, and the distinct histories of inter-communal relations in each region.

'Areas of relative affluence in NSW appear to be places where there is less racism. We believe it is education levels which are the most substantial contributor to this class-based geography of racism,' say Drs Kevin Dunn and Amy McDonald from the UNSW School of Geography.

'Contrary to popular belief, there are regions in Sydney where people harbour more racist sentiment than rural areas of NSW. The areas where the respondents were the most 'racist' were outer western Sydney, the mid north coast and the major industrial cities of the Illawarra and the Hunter.

'Intolerance of people with an Asian background is very high among both urban and rural Australians. Anti-Asian sentiment was not only widespread, but there appears to be little urban and rural variation,' says Dr Dunn. 'Anti-Indigenous feeling was higher among respondents from rural Australia. The anti-Indigenous sentiment that was found to exist in areas such as Richmond-Tweed and the north-western part of the state are likely to be due to the history of community relations in those areas.'

The study found that although a region demonstrated intolerance of a specific group this did not necessarily translate to high levels of opposition to multiculturalism. Also, the geographies of anti-Asian and anti-Indigenous sentiment do not match. In contrast to their anti-Indigenous sentiment, residents of the Richmond-Tweed and the Murrumbidgee areas were not overly concerned about multiculturalism.

'An urban-rural divide is only confirmed in so far as affluent Sydney, and to a lesser extent inner-western Sydney, are concerned,' says Dr Dunn. 'The survey generally shows that people in outer western Sydney, the Hunter and Illawarra are more intolerant of cultural difference than those in rural areas. Inner, eastern and lower northern Sydney were areas in which respondents exhibited lower levels of racism.'

Dr Dunn ascribes the variations in the different cultural make-up of each region of NSW, the different needs and resources of the cultural groups in each place and the different problems and tensions in each locality. 'This variation must be taken in to account when formulating anti-racism initiatives,' says Dr Dunn.

This pilot study feeds into a three-year project (2000–2003), funded by the Australian Research Council, which involves a survey of 10,000 Australians.

CONTACT DETAILS: Dr Kevin Dunn, School of Geography, tel. (02) 9385 5737 or

Victoria Collins, Public Affairs and Development, tel. (02) 9385 3644.
Date: 16 October 2000

Used with permission of UNSW Public Affairs.

Sending a media release

Once the release is written, what then? For large newspapers, send the release to the Chief of Staff as well as to the relevant rounds writer (for example, 'environment', 'economy', 'education', 'police' or 'local government'—but not all of them!). This may require that you do some background work to reveal which journalist does what. However, if you already have a working relationship with a specific journalist and it seems appropriate to send your release to that person, then do so, but first check that he or she is not unwell or on holiday as your story may otherwise go unattended. For small newspapers, send the release to the Editor. And for television and radio, send the release to the News Director or Chief of Staff. Do not make the mistake of sending your media release to the Chief Executive Officer, Managing Editor, Managing Director or someone else at 'the top', unless you know them personally. All this is likely to achieve is a delay, as the release filters its way down to the Chief of Staff.

Find out who you should send your media release to.

In general, it is probably best to email or fax a news release, although if you live in a town with a small local newspaper, or a radio or television station, there may be some advantage in delivering the story yourself. You may then get the opportunity to make your case personally and begin to develop a working relationship with a local journalist. If you fax a media release, try to keep it simple and confine your message to a single page. A release to a media outlet's central fax number is almost sure to be read by at least one person and that person will make decisions about what to do with release on the basis of a quick review. So be sure the media release is professionally written and presented. If you email your release, direct it to specific and relevant journalists and editors. Do not send your release to everyone in blanket email coverage. Unsolicited bulk messaging is not welcomed and is unlikely to achieve the ends you seek. If you do email your release, be sure the text is in the main body of the message, not sent as an attachment. Many people are reluctant to open unsolicited attachments. Of course, if you send images, these will need to be included as attachments, but the images should be identified or referred to in the text of your email and should be of high quality.

It is important to note that media staff change, offices relocate and media enterprises are bought and sold. As a result, your contacts at various outlets may change, so keep your media contacts file up to date. Also, no one is going to return your release with a 'return to sender' on it. It will go into the bin.

Following up a media release

You are unlikely to hear back immediately about your media release unless your story is especially important or engaging. However, if you do not hear back within a few days from the media outlet to which you sent your release, consider contacting them to see if the story is of any interest to them and to see if they need more information. This is also a good opportunity to convince the reporter that your story is worth covering. Try not to badger the poor journalist, as they may be reluctant to take your next call.

It is often difficult to foresee whether any particular story will get coverage. Some stories will be reported. Others will not. If your story did not make it, share your disappointment with the reporter, but do not get angry about it. That is unlikely to do your case any good—in either the short term or the long term.

Developing a rapport with journalists can be a mutually beneficial exercise. As well as getting the author's news or views into the community, regularly writing media releases can help to improve the author's writing style. The journalist, on the other hand, may gain a regular and productive source of interesting news stories.

If you are favoured by good media coverage, phone or write to the reporter and express your appreciation to them. Like anyone else, they will welcome positive feedback. If, on the other hand, there are errors or problems with your story, be courteous in any complaints you make. Remember, journalists are ordinary people whose mistakes are more public than most people's!

Being interviewed

After you have sent out a media release, you may be contacted by a reporter seeking follow-up information. This might be by phone. Although you might be a bit nervous about this, you should agree to speak—especially if you wish to develop an ongoing professional relationship with the journalist. However, do not allow the interview to proceed at a faster pace than you feel comfortable with. If you have been caught at a bad time (for example, dinner is burning) or you need a few moments to gather your thoughts and notes, tell the reporter that you will call them back in a few minutes. Use that short time to ensure you are clear about what you want to say, and then be sure to call the reporter back *when you said you would*. During the interview take time to explain your points, but try not to 'waffle on' or dwell on marginal issues. If you have discussed an issue that is critical to the story, ask the reporter to read back their notes to ensure that 'we've got it right'. You are unlikely to encounter an objection as it will help the journalist avoid disputes about material after the paper has been published.

Before you do a media interview, get clear in your mind the key messages you want to convey.

While it might be considered unethical in some academic work to quote material from interviews without letting the speaker preview the ways in which their words will be used, reporters will not normally let you see a copy of the story they plan to publish before it goes to subeditors or to print. Asking to scrutinise the story before publication is unlikely to achieve much other than upsetting the reporter.

Sometimes your release will find its way to a television or radio station. This kind of interview can be fairly unnerving and a few precautions are always worthwhile.

- If the interview is to be conducted over the phone, make sure that you are available to receive the call. If you are to be on a live show, you will have been scheduled in and other interviews cannot be rescheduled.
- Try to ensure that there are no barking dogs, rustling paper, unexpected visitors or mischievous friends to interrupt you (a DO NOT DISTURB sign can be very useful). Turn off any devices such as printers that may make noise during the interview. And try not to move your chair, tap your pen or drum your fingers on the desk during the interview.
- Have your press release and a note pad in front of you. Jot down the most important parts of your story so you can get straight to the point and not be drawn into a discussion of insignificant details. The prompts on your note pad might also include reminders to greet the interviewer and to refer

to her/him by name. Radio interviewers will appreciate the recognition as it confirms their importance to their audience. Remember to say thank you. These simple prompts can put you at ease. Many people unused to being interviewed in such an immediate atmosphere become nervous and can forget key issues and day-to-day courtesies. Those omissions can make you seem ill-versed in your subject matter, and cold and distant to a listening audience. You might also find your note pad useful for writing brief notes on questions that are asked.

- During the interview, when the interviewer repeats some facts (usually directly from your release), make a point of saying 'that's absolutely correct' or 'you've got straight to the main cause of the problem'. This again makes the interviewer feel comfortable and welcoming towards you.

If the radio interviewer and producer both like you then you may become a regular guest on their show. You are now well on your way to fame and perhaps even fortune!

Try to develop a good long-term relationship with journalists. They may eventually see you as an expert and seek you out for stories and comment.

If you do find yourself in a position where you are asked to give a radio or television interview—perhaps on a regular basis—Mathews (1981) provides some very helpful additional advice. There is also a growing number of websites that offer guidance. See, for example, Berger (1997) and Telg (2011).

The results of writing and sending media releases can be enlightening, both to you and to the general public. At worst you will have explored an alternative genre of writing. At best you will have achieved your bit of fame and sparked interest in, and knowledge about, your activities as a geographer or environmental scientist within the wider community. The various media are vital vehicles by which we can endeavour to communicate the results of our labours to audiences larger than the common group of markers, colleagues, families and friends. If we want larger audiences within the general public to value our work and other contributions to the community, it is essential that we develop and refine our skills in presenting information to and through the media.

REFERENCES AND FURTHER READING

Berger, J. 1997, 'Teaching media interview skills', in *Teaching Public Relations*, ed. T. Hunt, Public Relations Division of the Association for Education in Journalism and Mass Communication, viewed 3 June 201, <http://lamar.colostate.edu/~aejmcpr/27berger.htm>.

Gee, M. 2012, *Margaret Gee's Australian Media Guide*, Crown Content, Melbourne.
  This resource includes detailed entries on more than 20 000 media contacts and 2500 media outlets. It includes a wealth of media contact information such as telephone and fax numbers, email addresses, lists of key staff, advertising rates and program titles.

Hay, I. & Israel, M. 2001, '"Newsmaking geography": Communicating geography through the media', *Applied Geography,* vol. 21, no. 2, pp. 107–25.

Mathews, I. 1981, *How to Use the Media in Australia*, Fontana, Melbourne.

Minnis, J.H. & Pratt, C.B. 1995, 'Let's revisit the newsroom: What does a weekly newspaper print?', *Public Relations Quarterly*, Fall, pp. 13–18.

Mondo Code LLC 2011, *Mondo Times*, viewed 3 June 2011, <www.mondotimes.com/index.html>.
  A worldwide, searchable guide to print, audio and visual media.

Stewart, S.A. 2004, *Media Training 101: A Guide to Meeting the Press*, John Wiley, Hoboken, New Jersey.
  This helpful book offers key sections on understanding the media, facing the media, and how to give a great interview.

Telg, R. 2011, *Meeting the Press: A Comprehensive Media Relations Training Program*, viewed 3 June 2011, <http://aec.ifas.ufl.edu/mediarelations/index.html>.
  As its title suggests, this is an extensive and very helpful set of resources on dealing with various media.

Walters, T.N., Walters, L.M. & Starr, D.P. 1994, 'After the highwayman: Syntax and successful placement of press releases in newspapers', *Public Relations Review*, vol. 20, no. 4, pp. 345–56.

Williams, D. 1994, 'In defense of the (properly executed) press release', *Public Relations Quarterly*, Fall, pp. 5–7.

FIVE

Preparing a poster

One picture is worth ten thousand words.

Frederick R Barnard (1927)

Key topics

- Why make a poster?
- How to produce a poster
- What are poster markers looking for?
- Designing your poster

Posters are a useful way of presenting the results of research and other scholarly enterprise. They are an effective and swift means of presenting an idea or set of ideas, and are finding increasing use at professional conferences and other gatherings (Pechenik 2007, p. 244). Producing a good poster can be challenging, as you need to provide an effective combination of graphic and written communication. This chapter outlines some of the keys to poster production, including layout, visibility and the use of text and colour. While there is also a brief reference to the use of figures, extended discussions on the production of graphical devices and maps can be found in Chapters 6 and 7, respectively.

Why make a poster?

A poster presents an argument or explanation, summarises an issue or outlines the results of some piece of research in succinct visual form. Posters are especially good for promoting informal discussion and showing results that require more time for interpretation than is possible in, say, an oral presentation. However, they are not useful for reviewing past research or presenting the results of 'textual' research (Lethbridge 1991, p. 14).

Physically, a poster is a piece of stiff card or laminated paper about 90–100 centimetres × 60–75 centimetres in size (though some may be much larger) on which graphic materials, such as maps, charts and photos, are displayed and linked together by a small amount of text. There are many examples of good posters on the web, including Hess, Tosney and Liegel (2010) and Purrington (2011).

Assessors may ask you to prepare a poster for a variety of reasons. Posters can:

- add variety and new challenges to the course
- encourage the expression of complex information and ideas through careful combinations of text and graphics
- develop and test skills in graphic communication
- stimulate critical thought
- offer the prospect of encouraging student–staff interaction (for a discussion, see Hay & Thomas 1999; Howenstine et al. 1988; Knight & Parsons 2003; Vujakovic 1995).

Posters tend not to be assigned often or early in undergraduate courses, despite the great value such an exercise can have. Instead they are often reserved for upper-level and postgraduate work, where they are often prepared to present research findings to a professional conference or public meeting. Posters are a more challenging means of communicating information than you might think. They require you to express complex ideas with brevity and grace, and to balance text with graphics of high quality.

How to produce a poster

There are two main techniques for producing posters. Although both employ the same design principles set out in this chapter, the methods of construction are different. The first technique involves physically 'cutting and pasting' images, maps and text onto a large sheet of stiff card. These posters may or may not then be laminated. The second technique involves creating the poster in software such as Adobe Illustrator, CorelDRAW or Microsoft PowerPoint and printing it on a large-format printer. Typically, these lighter-weight posters are then laminated to enhance their durability. The former 'cut and paste' method is becoming less common as opportunities expand to prepare posters by the second method.

Well-produced 'cut and paste' posters can be very effective, but they almost invariably look less professional than their large-format printed counterparts. By virtue of their size and rigidity, they are also usually more difficult to transport than large-format posters, which can generally be rolled up easily and transported in a hard-sided map tube. So, if you have experience with relevant software and access to a large-format printer (through your university or one of the websites that offers large-format printing), you should think seriously about preparing a poster with these tools.

What are your poster markers looking for?

Poster production is not an explicit test of your artistic abilities, but if you have such aptitudes you should take advantage of them. A few central principles are critical to creating a successful poster, as discussed in Box 5.1.

Box 5.1
Principles
of poster
production

Successful posters have the following characteristics:
- *Attention-getting*—the poster should make a good first impression and grab the viewer's attention. Achieve this through layout, colour, title and other devices.
- *Brevity*—the poster should make its points quickly.
- *Coherence*—an effective poster makes a logical, unified statement requiring no further explanation. It should be intellectually accessible to the intended audience and must be capable of standing alone.
- *Direction*—the poster should have a clear trajectory through the subject matter. Keep the poster simple. Keep it focused. Overcomplicated posters discourage and confuse readers.
- *Evidence*—the poster should present an argument that is supported by accurate, referenced evidence.

Not surprisingly, these principles are very similar to the keys that unlock successful written and oral communication.

Posters are typically produced for a broader audience than just your lecturer and classmates. For example, you might be asked to produce a public information poster on strategies for making best use of green waste. If you have such an assignment, always consider your target audience and ensure that your language, ideas and choice of graphics suit that audience. Perhaps more commonly, though, you will be working on 'academic' posters. In addition to the principles of poster production as set out above, these may require that you:

- express a problem and resolve it
- argue or explain an issue
- evaluate evidence concerning a chosen topic (Howenstine et al. 1988, p. 144).

Academic posters are quite different from promotional posters.

Figure 5.1 Poster assessment schedule

Student Name:	Grade:	Assessed by:

The following is an itemised rating scale of various aspects of written assignment performance. Sections left blank are not relevant to the attached assignment. Some aspects are more important than others, so there is no formula connecting the scatter of ticks with the final grade for the assignment. Ticks in either of the two boxes left of centre mean that the statement is true to a greater (outer left) or lesser (inner left) extent. The same principle applies to the right-hand boxes. If you have any questions about the individual scales, comment, final grade or other aspects of this assignment, please see the assessor indicated above.

Quality of argument

Clear statement of question or relationship being investigated					Ambiguous or unclear statement of purpose
Poster fully addresses the question					Poster fails to address the question
Poster 'stands alone' requiring no additional explanation					Poster is difficult or impossible to comprehend without additional information
Logical/orderly explanation of the issue under investigation					Illogical/inadequate explanation
All components in presentation given appropriate level of attention					insufficient/ unbalanced treatment of components

Quality of evidence

| Argument well supported by evidence and examples | | | | | Inadequate supporting evidence or examples |

| Accurate presentation of evidence and examples | | | | | Incomplete or questionable evidence |

Use of supplementary material

| Effective use of figures, tables and other illustrative material | | | | | Illustrative material not used when needed or not discussed in text |

| Illustrations presented correctly | | | | | Illustrations presented incorrectly |

Poster appearance

| Poster carefully produced | | | | | Sloppy presentation |

| All text legible from 1.5 metres | | | | | Text illegible from 1.5 metres |

Sources/Referencing

| Adequate number of sources | | | | | Inadequate number of sources |

| Adequate acknowledgment of sources | | | | | Inadequate acknowledgment of sources |

| Correct and consistent referencing style | | | | | Incorrect or inconsistent referencing style |

| Reference list correctly presented | | | | | Errors or inconsistencies in reference list |

Assessor's comments

Your poster should reflect critical thinking rather than simply your capacity to describe some phenomenon. If you are given free choice of poster topic by your lecturer, you would be wise to begin by checking that the approach to the topic you have selected is appropriate. Make sure that you are satisfying the intellectual demands of the exercise as well as the graphic requirements. The poster assessment schedule in Figure 5.1 indicates the general criteria for assessing posters. You will probably find it productive to continually assess your poster against these criteria as you work on it. This should help you produce a high-quality piece of work.

While a completed poster should require no additional information for the viewer to be able to understand the content, there may be some occasions when your lecturer will ask you to stand by your poster and answer questions about it—as is the practice at academic and professional conferences.

A good poster should stand alone, requiring no further explanation.

Designing your poster

Layout and research

Because thoughtful composition and layout are such important parts of effective communication (Vujakovic 1995, p. 254), it is worth reserving a big chunk of time for the layout of your poster. Work that may have taken many hours to prepare can be ruined by an ill-conceived layout. It is also a good idea to design your poster at the same time as you are conducting the research associated with the exercise. Produce drafts or mock-ups of poster layouts before deciding on the final set-up. Discuss these with friends and perhaps with your lecturer, who will all bring their fresh, critical eyes to your work. You will probably find that this will both save you time and help you to make sense of the issue under discussion. For example, when designing the poster you will probably find weaknesses and gaps in the argument you are developing or in the relationships you are exploring. These discoveries should prompt you to undertake additional enquiry that will contribute to the production of a better project. Of course, this means that you cannot leave poster production to the night before the assignment is due, no matter how straightforward the project may initially appear!

Figure 5.2   Examples of different poster formats

(a)

Title and author		
Introduction	Results—Year A	Conclusions
Project's significance		Further research
Materials and methods	Results—Year B	References

(b)

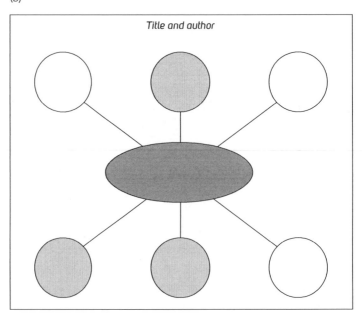

Source: Hay & Thomas (1999, p. 211).

Guiding your reader through poster components

Careful organisation is critical to good poster communication. Your poster should have a logical and clearly apparent structure—see Figure 5.2(a). To this end, it is useful to note that, apart from graphics, posters usually comprise six important components:

1   a *title and subheadings*, which should be meaningful, visible, brief and memorable. The title should normally also include the names and, where appropriate, the affiliations and contact details of the poster's authors.

2 an *abstract*, summarising the poster's key content. Some posters neither have nor need an abstract. Detailed information about writing abstracts is provided in Chapter 2.

3 an *introduction*, in which you provide a short statement of the problem investigated and the approach used.

4 the *body* of the presentation. If your poster is presenting the results of some research, this section might be split into a number of parts. The first is commonly *materials and methods*, where you explain any research techniques used in sufficient detail for readers to work out the scope of your study, the precision with which it has been done, and the validity of the data obtained. The second is likely to be a statement of *results*. This is a vital part of the poster. It enables readers to **examine** the data on which your conclusions are based and to critically evaluate their validity. Most readers need to see the information on which conclusions are based before they are able to accept them.

5 a *statement of conclusions* and/or *directions for future research*. The conclusions let readers know what interpretations you have made of the information presented in the poster, drawing together information from the introduction to highlight its significance. You might also explain how future work could follow up on some particular aspect of your findings.

6 *references*, which are placed on the front of the poster or, in some cases, attached to the reverse side.

You may also wish to add acknowledgements that recognise those people or organisations who have made some valuable contribution to the work.

Evidently, then, the structure of posters has much in common with essays and other forms of expression. However, unlike written forms of communication, posters do not have to be set up in a linear form, with readers moving from top left to bottom right. Two alternatives to the linear structure are 'spider' diagrams—see Figure 5.2(b)—which might show factors contributing to social power in a community, and cyclical diagrams, which could depict the hydrological cycle or the cycle of poverty. However, while posters do offer this flexibility of presentation, it is important that you provide your readers with a clear sense of direction. To this end, readers must usually be guided through the poster with informative subheadings, numbers or arrows. Readers will not know their way around your poster as you do. Lead them along. Be aware, however, that after reading the title and scanning the poster, many people move straight to the conclusions, before working out whether to read the poster more thoroughly or move on to another. If the conclusions seem important, the viewer may decide to read the rest of the poster. Give some hard thought to your title and to the way you present your conclusions.

Make sure your poster's structure is clear to readers.

Posters may also be set up to be interactive (Vujakovic 1995, p. 254). Readers might be asked questions to which the answers are revealed by lifting flaps or overlays, or by turning circles of cardboard. Devices such as these will almost certainly encourage your audience to interact with the poster. They should be both uncomplicated and robust to reduce the chances of the poster falling apart as it is used.

Text

Number of words and level of detail

Use small amounts of text on your poster and be sure it is visible from at least 1.5 metres.

Unlike advertising and promotional posters, academic posters almost always contain some amount of text. Together with the graphics, text contributes to the introduction, explanation and discussion of your work. It is important, however, to keep text to a minimum. Lethbridge (1991, p. 18) suggests a maximum word count of 500, but makes the point that a poster will be much more effective if it uses only 250–400 words. Confine the text you do use to short sections that complement the poster's graphic components. Make sure no single block of text takes longer than 30 seconds to read (Knight & Parsons 2003, p. 155). Poster viewers do not enjoy reading long texts. Bad posters are often bad because they contain too much text or because all the text is presented in a small number of blocks.

You do not need to provide all the intricate details of a project discussed in a poster. In much the same way as a talk highlights points that can later be explored by interested members of the audience, so a poster can be used to present the most important elements of your work. Onlookers whose interest has been aroused might then speak to you about details. That said, an earlier point bears repetition: the poster must still be comprehensible without further elaboration.

Visibility of text

Given the amount of work required to get the number of words in your poster down to an acceptable level, doubtless you will want your audience to be able to read each carefully selected letter! To this end:

- Ensure that all materials on your poster are clearly legible from a distance of about 1.5 metres, and that titles and headings are discernible to viewers several metres away.
- Use upper and lower case letters throughout headings and poster text. Do not just use capitals. THEY ARE A LOT MORE DIFFICULT TO READ QUICKLY.

- Confine textual material to a number of brief statements. Do not write an essay and paste it to the board! Few people, if any, will read it.
- Use larger fonts and/or display faces for short headings, and smaller and/or less decorative type for longer titles (see Box 5.2). Twelve point—a common size for typed essays and reports—is much too small a point size for the body of the text. If you need to use 10–12 point type to fit all your text on the poster, the poster probably contains too much text. The solution to this problem is to abbreviate and rewrite the text.

For posters, suggested type sizes are as follows.

- main headings: 96–180 point (27–48.5 millimetres)
- secondary headings: 48–84 point (12.9–25.4 millimetres)
- section headings: 24–36 point (5.9–8.7 millimetres)
- text and captions: 14–18 point (3.2–4.6 millimetres)

**Box 5.2
Suggested poster type sizes**

Type is one of the most important aspects of visual design, especially for headings, and you should use a typeface that relates to the subject material. While the body of the text needs to be clear and simple, you can add extra graphic character to the poster through your choice of a suitable type. Typefaces and type styles (as discussed in Box 5.3) play important roles in attracting attention, and contribute to the poster's overall theme and clarity.

- The *typeface* is the particular character of the letter forms, from which there are thousands to choose (for example, **Helvetica** and Times).
- *Type weight* is the thickness of the letter stroke (for example, regular and **bold**).
- *Type style* may be roman (upright), *italic* (slanted), condensed or **extended**.
- *Type size* may be 14 point, 16 point or any size necessary.
- *Type colour* may be one-colour (black), two-colour (black with a second spot colour) or full-colour. Even with one-colour printing, text may be black, white or shaded.

**Box 5.3
Characteristics of typefaces and type styles**

For example, a very modern typeface might be appropriate for a poster investigating some of the effects of space-adjusting technology on international trade practices and patterns, while a playbill typeface—first used in the Victorian era on posters advertising stage productions—could be appropriate for a poster considering nineteenth-century Australian health care conditions.

Colour

One of the most striking and emotive elements of many posters is colour, a component that adds to or detracts from the overall impact of the project. Colour can command attention, bring pleasure and clarify a point (Larsgaard 1978, p. 193). It can also highlight important dimensions of a poster or suppress less important facets.

Use type and colour that complement your poster.

Although colour is important, it is important to use it judiciously. To avoid confusion and chaos in the poster, use as few colours as possible. Pechenik (2007, p. 249) suggests that a single background colour should be used to unify the presentation. Also take care with your choice of combinations of text and background colours. There should always be enough contrast between text and background to allow the text to be read easily from a distance. For example, orange text on a yellow background can be difficult to read, as can red text on a green background. If you set text over an image used as background to your poster (for example, a watery background for a poster on environmental consequences of dams), make sure there is sufficient contrast between the image and all of the text to allow someone to read your message easily.

Colour can add symbolic connotations and feelings to the message of a poster. You might find the list in Box 5.4 useful. Adapted from Sim (1981), it summarises some colour-connotation connections from a white Anglo-American perspective.

**Box 5.4
Some colour connotations**

• black	clear-cut and crisp, death, dignity, doom and gloom, financial credit, formality
• blue	calm, climatological, coastal, coolness, rivers, peace, sadness
• brown	dismal, dreary, earth, pollution, soils
• green	agriculture, conservation, coolness, envy, freshness, growth, nature, rural, safety, spring, vegetation, wealth
• orange	autumn, flames, healthiness, sunshine, warmth
• red	action, blood, danger, financial deficit, fire, hazards, health, heat, Marxism, noise, passion
• white	cleanliness, glory, iciness, purity, snow
• yellow	beaches, happiness, light-heartedness, sand, sunshine, weakness

Colour also may be used to add information to particular graphics. For example, the appropriate use of red and black conveys the message that financial results represent profit and loss statistics, not simply dollar figures.

Tables, figures and photos

Tables

Tables used on posters should generally be kept simple and clear. Readers are unlikely to spend much time trying to decipher complex data sets. In many cases, it is better to summarise and depict tabular information in the form of histograms, pie charts and other graphic devices. These will usually communicate your messages faster and more memorably, and for some audiences may be the most appropriate communication device. Although tables are a very useful means of communicating precise numerical information, ask yourself whether the data could be transformed into a 'picture' as opposed to a table. For example, a table of population growth figures for Rarotonga since 1900 might be better presented as a line graph. Try to achieve a balance between the extent to which you use tables and figures in your poster.

Figures and photos

In the process of considering the production of figures for a poster, you should address several questions. These include:

- What type of 'picture' will best illustrate the point? For example, would a pie chart be more effective than a bar chart?
- What symbols and colours can you employ to make a greater impression and to communicate the idea more clearly? For example, would it be effective to illustrate South Africa's balance of payments history in a bar graph depicting piles of coins shaded red or black depending on each year's deficit or surplus? The graphical devices you employ may be more readily understood, and may therefore be more effective, if they incorporate symbols and colours that have evolved through tradition, convention and public recognition to be representative of their content.
- Are all the graphic conventions likely to be comprehensible to the audience? Maps, for example, should always have a legend or key that explains all the symbols used.

Consider using a figure to summarise data for your poster.

Figure 5.3 shows examples of statistics from the same research that might be incorporated into a poster display in different forms. In general, it is advisable to find a simplified but graphically interesting way of displaying data when producing a poster. Whereas the table might provide the most accurate record of research data collected, it does not enable the viewer to absorb the major trends as quickly as a graphic depiction does. A detailed graph, which might be used in a written report, would enable the reader not only to obtain a picture of the relative trends quickly but also to make a reasonable interpretation of the data. However, a graph such as this is still unnecessarily detailed for a poster display. A simplified and more interesting portrayal, like the third item in Figure 5.3, enables the viewer to grasp the essential pattern quickly. In this example of a graph suitable for a poster display, interest and clarity are achieved by differentiating linework, type style and size. The addition of colour might increase visual appeal. Do not forget to include an appropriate title and, if necessary, a caption for figures.

Figure 5.3 Converting a numerical table into a poster-ready figure

	COMPANY EXPENDITURE		
YEAR	LOCAL	OVERSEAS	TOTAL
1981	5,067,000	1,010,000	$6,077,000
1982	5,328,000	3,126,000	$8,454,000
1983	6,141,000	2,842,000	$8,983,000
1984	7,002,000	6,989,000	$13,991,000
1985	8,269,000	2,354,000	$10,623,000
1986	2,103,000	2,413,000	$4,516,000
1987	2,025,000	4,201,000	$6,226,000
1988	1,567,000	3,368,000	$4,835,000
1989	4,824,000	423,000	$5,247,000
1990	6,041,000	0	$6,041,000

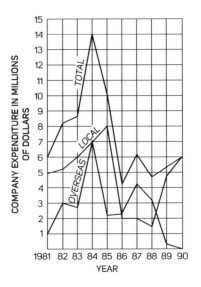

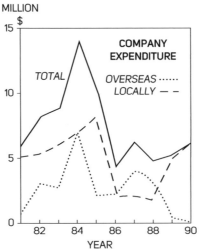

Images are a particularly important part of a poster. For example, you can use photographs to show pasture before and after the eradication of rabbits, or to illustrate transformations in a city after a large industry closes local operations and moves overseas. You can use accompanying diagrams to explain the processes and procedures underlying those surface appearances (Vujakovic 1995, p. 253). If you use photographs in your poster, be sure they are of high quality (in focus and with sharp contrast) and large enough to be clearly visible from 1–2 metres. If the size of the photographed object would not be immediately clear to your audience, provide some idea of **scale**; for example, by including your camera lens cover in the photograph. Do not forget to provide an appropriate title for the photograph (and for other figures on the poster).

The web offers a vast range of helpful images, many of which are stock images that can be used freely. Others can be used subject to copyright restrictions, which are often very generous if the image is used for academic purposes (as opposed to a money-making concern). In addition to facilities such as Google's Image search capability, there are also growing numbers of specialist online image galleries that can prove helpful (for example, University of Chicago Library's American Environmental Photographs Collection, the Royal Geographical Society with the Institute of British Geographers' Picture Library, or the National Geographic website). Given its importance, it is vital that graphic material in your poster be bold and relevant. Avoid filling up your work with pleasing but unnecessary pictures as these are likely to detract from the central matters you wish to communicate.

The web provides a vast array of images useful for posters.

You will work long and hard to prepare a good poster. Protect your investment. Consider getting it laminated, especially if you plan to use it more than once.

If you want to keep your poster, present it at a conference or display it permanently, consider getting it laminated.

Acknowledging sources

As with any other form of academic communication, you must accurately acknowledge any information and ideas, figures and facts, or text and tables you have drawn from other sources. These should be cited in exactly the same way as they would be in an essay or report (see Chapter 10 for a discussion of referencing conventions). In some cases, and particularly for formal academic posters, it is appropriate to include your reference list on the front of the poster. To save space, this might be in a smaller font than the poster's other text. In other situations (for example, community information posters) it may be more appropriate to firmly affix the list of references to the back of the poster, where they do not detract from the main message of your work. If you are at all uncertain about the correct location for references, speak to your lecturer.

A final note ...

Throughout the production process of your poster, keep in mind the principles that were introduced earlier: attention-getting, brevity, coherence, direction and careful use of evidence. Providing, of course, that the poster draws on good research, careful application of these principles should contribute to the production of first-class work.

REFERENCES AND FURTHER READING

Brown, B.S. 1996, 'Communicate your science! ... Producing punchy posters', *Trends in Cell Biology*, vol. 6, pp. 37–9.

Eggart, M.L. undated, *Effective Poster Design for Academic Conferences*, viewed 31 May 2011, <www.ga.lsu.edu/Effective%20Poster%20Design%20for%20Academic%20Conferences.pdf>.
A very useful slide show prepared by a cartographer setting out illustrated points on effective poster design.

Hay, I. & Thomas, S. 1999, 'Making sense with posters in biological science education', *Journal of Biological Education*, vol. 33, no. 4, pp. 209–14.

Hess, G.R., Tosney, K. & Liegel, L. 2010. *Creating Effective Poster Presentations*, viewed 31 May 2011, <www.ncsu.edu/project/posters>.

Howenstine, E., Hay, I., Delaney, E., Bell, J., Norris, F., Whelan, A., Pirani, M., Chow, T. & Ross, A. 1988, 'Using a poster exercise in an introductory geography course', *Journal of Geography in Higher Education*, vol. 12, no. 2, pp. 139–47.

Knight, P. & Parsons, T. 2003, *How to Do Your Essays, Exams, and Coursework in Geography and Related Disciplines*, Nelson Thornes, Cheltenham.

Larsgaard, M. 1978, *Map Librarianship*, Libraries Unlimited, Littleton, Colorado.
Includes discussion of some useful design principles.

Lethbridge, R. 1991, *Techniques for Successful Seminars and Poster Presentations*, Longman Cheshire, Melbourne.
This short book briefly introduces poster design principles and goes on to provide a comprehensive review of technical aspects of preparing illustrations. Indeed, the book's title may be a little misleading given the volume's overall emphasis on graphic production techniques.

Matchett, S. 2009, *Designing an Academic Poster*, viewed 31 May 2011, <https://www. wiki.ed.ac.uk/display/HowTowiki/Designing+an+academic+poster>.

A helpful four-part online guide to good poster design and production.

Nicol, A.A.M. & Pexman, P.M. 2010, *Displaying Your Findings: A Practical Guide for Creating Figures, Posters, and Presentations*, 6th edn, American Psychological Association.

An extensive and very successful volume with a helpful chapter (13) on poster preparation.

Pechenik, J.A. 2007, *A short guide to writing about biology*, 6th edn, Pearson Longman, New York.

Chapter 12, 'Writing a poster presentation', offers a useful and detailed discussion on the process of converting a written paper into a good poster presentation.

Purrington, C. 2011, *Advice on Designing Scientific Posters*, viewed 31 May 2011, <www.swarthmore.edu/NatSci/cpurrin1/posteradvice.htm>.

Sim, R. 1981, *Lettering for Signs, Projects, Posters, Displays*, 2nd edn, Learning Publications, Balgowlah.

University of Guelph Teaching Support Services, undated, *Effective Poster Design*, viewed 31 May 2011, <www.soe.uoguelph.ca/webfiles/agalvez/poster/>.

As well as a comprehensive overview of poster preparation, design and construction, this site also includes examples of student posters and an assessment schedule.

Vujakovic, P. 1995, 'Making posters', *Journal of Geography in Higher Education*, vol. 19, no. 2, pp. 251–6.

Readable and concise discussion of poster production.

SIX

Communicating with figures and tables

As Bertrand Russell once said, most of us can recognise a sparrow, but we'd be hard put to describe its characteristics clearly enough for someone else to recognise one. Far more sensible to show them a picture.

Rowntree (1990, p. 193)

Key topics

- Why communicate graphically?
- General guidelines for clear graphic communication
- Different types of graphic
- Preparing a good table

In geography and the environmental sciences, words or numbers alone are often not sufficient to communicate information effectively. Graphic communication allows you to succinctly display a large amount of information and helps your audience to readily absorb it. Effective illustrations can help a reader achieve a rapid understanding of an argument or issue.

Why communicate graphically?

Figures employ human powers of visual perception and pattern recognition, which are much better developed than our capacity to uncover meaningful relations in numerical lists (Krohn 1991, p. 188). Often we see things in graphic form that are not apparent in tables and text. Krohn argues that graphs reveal relationships that allow both the numbers upon which they are based and the concepts by which we understand those numbers to be reinterpreted. As such, graphs are critical interactive sites for comprehending the world around us.

Graphic material can reveal relationships difficult to see in tables and text.

In addition to their intellectual functions, graphics can enhance various forms of technical writing in different ways (Eisenberg 1992, p. 81):

- *In essays and reports*, graphs and tables summarise quantitative information, freeing up text for comments on important features.
- *In instructions*, graphics may help people to understand the principles behind the operation of some process or the characteristics of a phenomenon.
- *In oral presentations*, charts, tables and figures relieve monotony, help to guide the speaker, and aid the audience's understanding of data.

This chapter discusses the character and construction of different types of illustrative material. It is important first to introduce a few general guidelines.

General guidelines for clear graphic communication

Good graphics are concise, comprehensible, independent and referenced.

Graphics are concise

Graphics should present only that information that is relevant to your work and required to make your point. Review the data you are going to portray to find out what they 'say' and then let them say it graphically with the minimum of embellishment (Wainer 1984, p. 147). If you reproduce an illustration or table you have found in your research, you may need to redraw or rewrite it to remove irrelevant details, while remembering to cite the original source.

Graphics are comprehensible

A good graphic should be comprehensible in its own right, requiring no additional explanation to be understood.

Your audience must understand what the graphic is about. Provide a clear and complete title that answers 'what', 'where' and 'when' questions, and include effective labelling. Effective data labels and axis labels are:

- easy to find
- legible
- easily associated with the axis or object depicted (they should be close together)
- readable from a single viewpoint; that is, a reader looking at the graphic should be able to read the text without having to turn the page sideways.

Although you should fully label your graphics, do make sure that **data regions** (that is, the parts of the graph in which the data is displayed) are as clear of notes, axis markers and keys as practicable. In short, the graphed information should be clear and easy to read.

If your graph displays two or more data sets, they must be easily distinguished from one another. Graphs should include no more than *four* simultaneous symbols, values or lines (Cartography Specialty Group of the Association of American Geographers 1995, p. 5) and each line or symbol should be different enough from the others to be distinguished easily.

You can also make a graphic more comprehensible by making effective use of the data region. Choose a range of axis scale marks that will allow the full range of data to be included while ensuring that the scale allows the data to fill up as much of the data region as possible. For example, if you are graphing data that range in value from 55 to 100, showing a scale that ranges from 0 to 100 would leave a great deal of empty space. If you take photographs, this principle will be familiar to you. Just as good photos will usually 'fill the frame', so a good graph will typically fill the data region. Finally, tick marks on each axis should also be placed at sufficiently frequent intervals for a reader to work out accurately the value of each data point (Pechenik 2009).

Graphics are independent

Graphics should stand alone. Someone who has not read the document associated with the graphic should be able to look at the table or illustration and understand what it means.

Graphics are referenced

You must acknowledge sources. Use an accepted referencing system to note sources of data and graphics (see Chapter 10 for information on referencing systems). Each graphic should be accompanied by summary bibliographic details (author, date and page in the case of the **author–date system**) or a **note identifier** allowing the reader to find out where the graph or the data upon which it is based came from. A reference list at the end of your work should provide the full bibliographic details of all sources.

Make sure that references are to the source *you* used, and not that of the author of the text you are borrowing from. For example, imagine you are copying a penguin population graph you found in a book written by Dr David De Bary and published in 2012. De Bary had, in turn, cited the source of his graphed data as the Argentine Penguin Research Foundation. Following the author–date system, the graph you present in your work would be referenced as (Argentine Penguin Research Foundation, in De Bary 2012, p. 12). The reference would not simply be to De Bary. Of course, there would also be a full reference to De Bary's work in the list at the end of your paper. Again, see Chapter 10 for further information.

Over and above these general principles, and unless there are good reasons to the contrary, graphs should include the following critical components:

- *number*—each graph should have a unique number (for example, Figure 4) allowing it to be identified in textual discussion.
- *title*—the title, which is placed above the graph itself, should be both brief yet precise enough to allow any reader to understand the information presented without reference to other texts. The title should answer 'what', 'where' and 'when' questions. If the title seems too long, consider adding a subtitle.
- *axis labels*—each of the graph's axes should include a succinct and accurate description of the item depicted on that axis, together with an indication of its units of measurement or magnitude where that may be required; for example, precipitation (mm) or population ('000).
- *data source*—an indication of the source of the graphed data should be provided. If you have copied the graph from somewhere else, an accepted form of citation must be used.

Different types of graphic

Various forms of graphic communication are described in this chapter, and some advice on their construction is given. In a deliberate strategy, all graphs in this book have been drawn using Microsoft Excel. While other, more powerful, software packages for producing graphics exist, Excel is a commonly available package that produces adequate figures for most undergraduate assignments. It is readily available in most universities, and students familiar with computers should be able to produce graphics comparable with (or better than) any of those shown in this chapter. Hand-drawn figures can easily be produced to the same standard. Irrespective of the technology used to produce graphs, it is important that you know what kind of graph is most appropriate for your data (and perhaps for your audience) and that you have a good sense of the conventions that underpin good graph design and production. This chapter offers some background to these key matters. Table 6.1 provides a summary of the major forms of graphic discussed, together with their nature and function.

Table 6.1 Types of graphic and their nature or function

Type of graphic	Nature or function
Scattergram (or scatter plot)	Graphic of point data plotted by (x,y) coordinates. Usually created to provide visual impression of direction and strength of relationship between variables.
Line graph	Values of observed phenomena are connected by lines. Used to illustrate change over time.
Bar chart	Observed values are depicted by one or more horizontal or vertical bars whose length is proportional to value(s) represented.
Histogram	Similar to bar graph, but commonly used to depict distribution of a continuous variable. Bar area is proportional to value represented. Thus, if class intervals depicted are of different sizes, the column areas will reflect this.
Population pyramid	Form of histogram showing the number or percentage of a population in different age groups of the total population.
Pie (circle) chart	Circular-shaped graph in which proportions of some total sum (the whole 'pie') are depicted as 'slices'. The area of each 'slice' is directly proportional to the size of the variable portrayed.
Logarithmic graph (log–log and semi-log)	Form of graph using logarithmic graph paper. Key intervals on logarithmic axes are exponents of 10. Log graphs allow depiction of wide data ranges.
Table	Systematically arranged list of facts or numbers, usually set out in rows and columns. Presents summary data or information in orderly, unified fashion.

Scattergrams

A **scattergram** or scatter plot is a graph of point data plotted by (x,y) coordinates (see Figure 6.1 for an example). Scattergrams are usually created to provide a visual impression of the direction and strength of a relationship between variables.

Figure 6.1 Example of a scattergram: *Life expectancy and total fertility rates, selected countries, 2009–10*

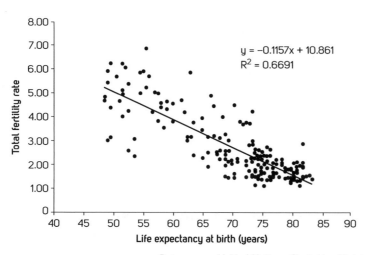

Data source: United Nations Statistics Division (2011).

Constructing a scattergram

The **independent variable** (that is, the one that causes change) is depicted on the horizontal x-axis and the **dependent variable** (the variable that changes as a result of change in the independent variable) is plotted on the y-axis. To illustrate the difference between independent and dependent variables, consider the relationship between precipitation levels and costs associated with flooding. Damage costs associated with flooding will usually depend on the amount of rainfall. Thus, rainfall is the independent variable (x-axis) and damage costs are dependent (y-axis). Or, the severity of injuries associated with a motor vehicle accident (dependent variable, y-axis) tends to increase with motor vehicle speed (independent variable, x-axis).

After points are located on the scattergram, you might draw a '**line of best fit**' (or **trend line**) through the points by eye (that is, your visual impression of the relationship expressed in the form of a line through the points). This line may be calculated mathematically and the regression equation expressed

Be sure to put independent and dependent variables on the correct axes of your graph.

on the graph (see Figure 6.1). It is important to note that a line of best fit is intended to depict the data *trend*—not to connect the dots on the scattergram.

Line graphs

Typically, **line graphs** are used to illustrate continuous changes in some phenomenon over time, with any trends being shown by the rise and fall of the line. Line graphs may also show the relationship between two sets of data. Figures 6.2, 6.3 and 6.4 are examples of line graphs.

Do not use a line graph if you are dealing with disconnected data (Eisenberg 1992, p. 97). For example, if you have air pollution data for every second year since 1945, the information should be graphed using a bar chart because a line graph would incorrectly suggest that you have data for each intervening year.

Figure 6.2 Example of a line graph: *Number of sites inscribed on the World Heritage List, 2004–11*

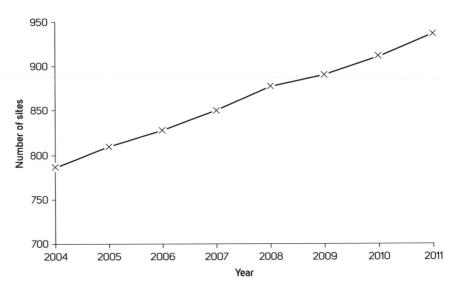

Data source: UNESCO (2011).

Constructing a line graph

Plot each (x,y) data point for your data set(s). When all the data points are plotted, join the points associated with each data set to produce lines such as those shown in Figures 6.2, 6.3 and 6.4. In some cases, however, it is more appropriate to draw smooth curves than it is to 'join the dots' (Mohan, McGregor & Strano 1992, p. 284; Pechenik 2009, p. 179) where, for example, a clear trend is disrupted by a single inconsistent data point. Make any such judgments very carefully.

Figure 6.3 Example of a line graph: *Permanent arrivals and departures, Australia, 1959–64*

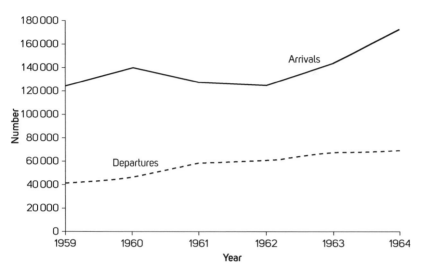

Data source: Commonwealth Bureau of Census and Statistics (1965).

Figure 6.4 Example of a line graph: *Mean NDVI following application of three different herbicides*

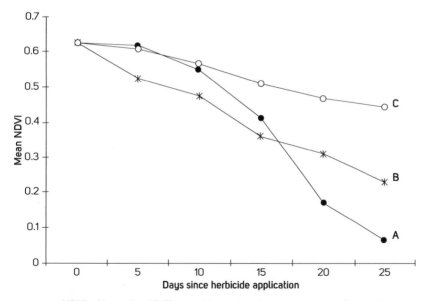

NDVI = Normalised Difference Vegetation Index, a measure of vegetation 'vigour'.

Figures calculated using red and NIR (near-infrared) reflectance recordings from 30 × 1 metre square test plots (ten plots for each herbicide).

If a number of lines are depicted in one graph, ensure that they can easily be distinguished from one another (see Figures 6.3 and 6.4) through use of colour, dotted lines or labels.

If you are showing average (mean) values on your graph, you can usefully provide a visual summary of the variation within the data by, for example, depicting the data range or the standard deviation about the mean (Pechenik 2009, p. 180). An example is shown in Figure 6.5.

Figure 6.5   Example of a graph showing variation within data (mean and range): *Mean (including trend line) and range (minimum–maximum) of NDVI measurements.*

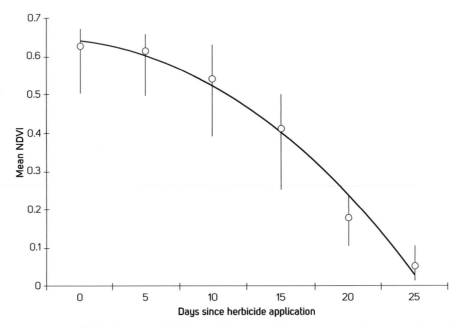

NDVI = Normalised Difference Vegetation Index, a measure of vegetation 'vigour'.

Figures calculated using red and NIR (near-infrared) reflectance recordings from 30 × 1 metre square test plots (ten plots for each herbicide).

Pechenik (2009, p. 181) offers additional helpful advice on representing variations within data, pointing to the utility of depicting both range and standard deviation:

> Plots of standard deviations or standard errors are always symmetrical about the mean and so convey only partial information about the range of values obtained. If more of your individual values are above the mean than below the mean, the error bars will give a misleading impression about how the data are

actually distributed. If your graph is fairly simple, you may be able to achieve the best of both worlds, indicating both the range and standard deviation (or standard error).

An example is provided in Figure 6.6.

Figure 6.6 Example of a graph showing variation within data (data range, mean and standard deviation): *Variations in NDVI measurements.*

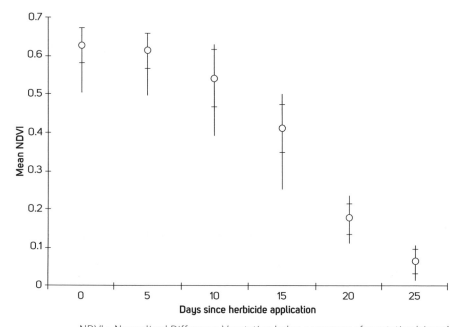

NDVI = Normalised Difference Vegetation Index, a measure of vegetation 'vigour'.

Figures calculated using red and NIR (near-infrared) reflectance recordings from 30 × 1 metre square test plots (ten plots for each herbicide).

If you do include indicators of variation in your graph, make sure that you include in your caption of notes that accompany the figure details of what you have plotted, together with the number of measurements associated with each mean (Pechenik 2009, p. 182).

Line graphs will sometimes compare things that have different measurements. This can be done by using vertical axes on the left and right sides of the graph to depict the different scales. Figure 6.7 provides an illustration of the use of multiple vertical axis labels. It is also an example of a **climograph**, a commonly used graph with two axes: one depicting temperature and the other showing precipitation.

Figure 6.7 Example of a graph using multiple vertical axis labels: *Darwin climate, average monthly rainfall and daily maximum temperature, 1941–2004*

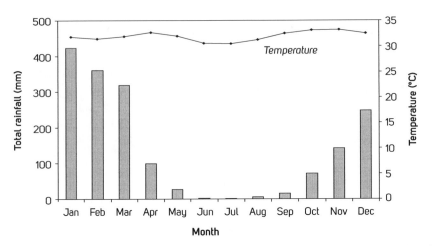

Data source: Bureau of Meteorology (2004).

Bar charts

Bar charts are of two main types: *horizontal* and *vertical*. Figures 6.8 and 6.9 show each type, respectively. Horizontal **bar graphs** usually represent a single period of time, whereas column graphs may represent similar items at different times (Moorhouse 1974, p. 67).

Figure 6.8 Example of a horizontal bar graph: *Queensland's rural water use for agricultural commodity production, 2008–09*

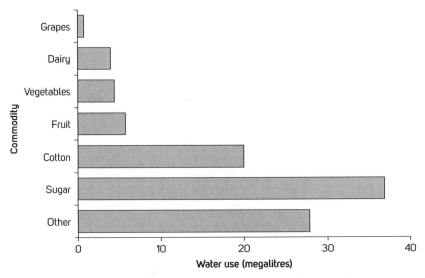

Data source: Queensland Department of Environment and Resource Management (2011).

Figure 6.9 Example of a vertical bar graph: *Total dwelling units approved, South Australia, July 2010 – June 2011*

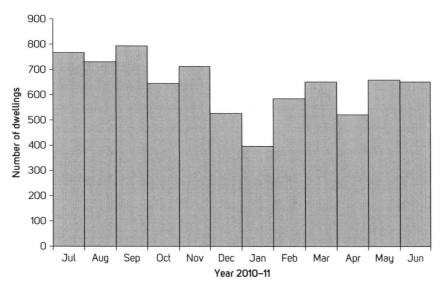

Data source: Australian Bureau of Statistics (2011).

In bar graphs, the *length* of each bar is proportional to the value it represents (Coggins & Hefford 1973, p. 66). It is in this regard that bar graphs differ from histograms, with which they are sometimes confused. Histograms use bars whose *areas* are proportional to the value depicted. See the following section for more on histograms.

Values represented are proportional to bar length in the case of a bar graph and to bar areas in the case of a histogram.

Bar charts are a commonly used and easily understood way of taking a snapshot of variables at one point in time, depicting data in groups, and showing the size of each group (Moorhouse 1974, p. 64; Windschuttle & Windschuttle 1988, p. 278). Figure 6.10 achieves all of these ends in a single graph.

Bar charts can also be used to show the components of data as well as data totals. See Figure 6.11 for an example of such a *subdivided bar chart*. It is possible to go one step further and represent data in the form of a *subdivided 100% bar chart* (see Figure 6.12 for an example). These can be useful for depicting figures whose totals are so different that it would be almost impossible to chart them in absolute amounts (Moorhouse 1974, p. 66).

Figure 6.10 Example of a bar chart: *Life expectancy at birth by sex and state/territory, Australia, 2004–06*

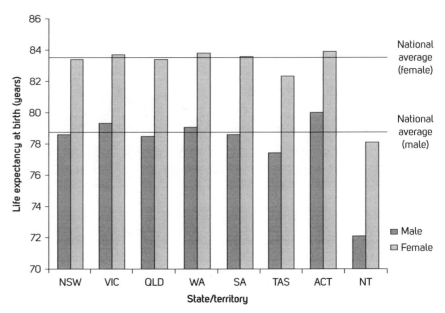

Data source: Australian Institute of Health and Welfare (2011).

Figure 6.11 Example of a subdivided bar chart: *Comparison of part-time and full-time employment in Australia, 1985–2005*

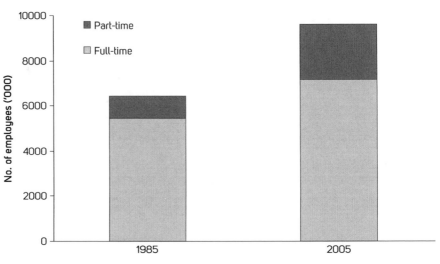

Data source: Australian Bureau of Statistics (2005).

Figure 6.12 Example of a subdivided 100% bar chart: *Household type by major Australian cities, 2001*

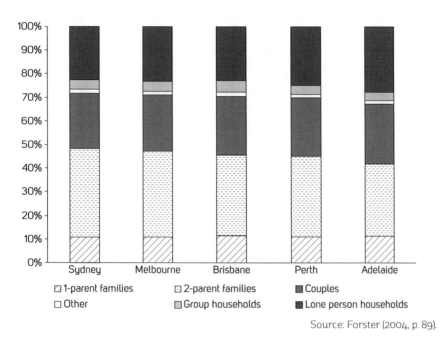

Source: Forster (2004, p. 89).

As Figure 6.13 illustrates, bar charts also can be used to portray negative as well as positive quantities.

Figure 6.13 Example of a bar chart depicting positive and negative values: *Annual population growth rates for selected countries, 2000–05*

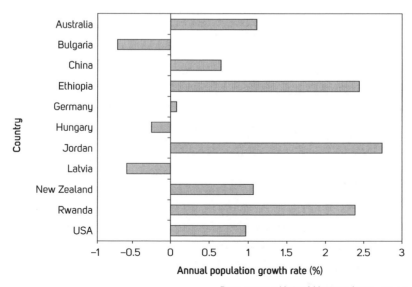

Data source: United Nations (2004, pp. 45–50).

Constructing a bar chart

Examine the data that is to be graphed and select suitable scales for the graph's axes. In general, scales should begin at zero (Coggins & Hefford 1973, p. 66) although this is not critical. Label the axes, being sure to label zero if it is included on a scale.

When the chart is being designed, it is important to consider the sequence of items being depicted. In general, the items should be listed in order of importance to the viewer. However, in simple comparisons in a horizontal bar graph format, it is best to arrange the bars in ascending order of length from bottom to top. That said, you must also be aware that some data sets are listed, by convention, in particular orders. For example, in Australia, Bureau of Statistics occupational groups are typically listed in the following order: Managers & Administrators, Professionals, Para-professionals, Tradespersons ... Labourers & related workers, Inadequately described, Not stated. Similarly, industrial groups are usually listed in order through primary (for example, farming), secondary (for example, manufacturing), tertiary (for example, retail) and quaternary (for example, information transfer) divisions. Graphs should typically reflect such customary presentation forms. If you are not sure whether to set up a table in ascending order, speak to your lecturer.

The next step is to draw in the bars. Their width is a matter of choice, but should be constant within a graph. If you use different widths within the same graph, some readers may be led to believe that bar width, and hence area, is more important than lengths. Bars should be separated from one another, reflecting the discrete nature of the observed values and the space between the bars should be about one-half to three-quarters of their width. However, where the pattern of change is of greater importance than the individual values, no space between the observations is left at all (Coggins & Hefford 1973, p. 66). For example, compare Figures 6.9 and 6.10. In Figure 6.9 the pattern of change through the year is of more interest to the reader than is the specific value for each month. By contrast, the different (and therefore graphically differentiated) life expectancies in each state/territory portrayed in Figure 6.10 are central to the chart's message.

Make sure your graph is labelled fully and correctly.

Finally, add appropriate title, labels, key and reference.

Histograms

Histograms are mainly used to show the distribution of values of a continuous variable. A continuous variable is one that could have any conceivable value within an observed range (for example, plant height, rainfall measurements or temperature), including fractional values, and may be contrasted therefore with **discrete data** in which no fractional numbers such as halves or quarters

exist (for example, plant and animal numbers). For examples of histograms, see Figure 6.14. This figure shows three histograms drawn using the same data set but different class intervals. Class intervals are explained shortly.

Figure 6.14   Example of histograms drawn using the same data but different class intervals

(a) *Hotel accommodation costs, Wellington, 1994*

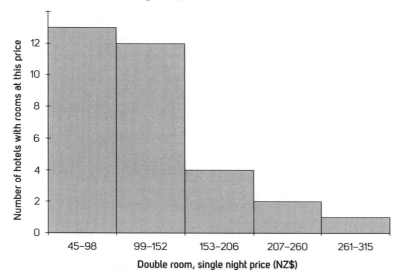

(b) *Hotel accommodation costs, Wellington, 1994*

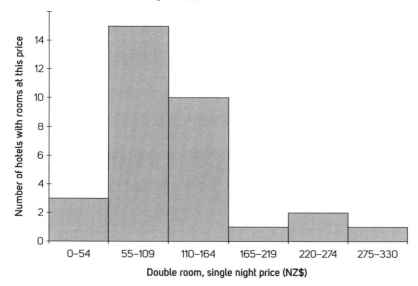

(c) *Hotel accommodation costs, Wellington, 1994*

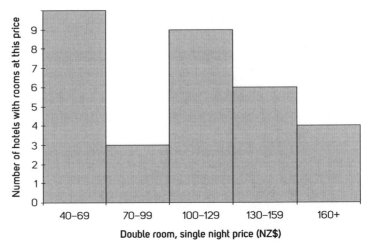

Data source: New Zealand Tourism Board (1995).

Histograms may be confused with vertical bar charts or column graphs, but there is a technical difference. Strictly speaking, histograms depict frequency through the *area* of the column, whereas in a column graph frequency is measured by column *height*. Thus, while histograms usually have bars of equal width, if the class intervals are of different sizes the columns should reflect this. For example, if one class interval on a graph was $0 to $9 and the second was $10 to $29, the second should be drawn twice as wide as the first.

The phenomenon whose size is being depicted is plotted on the horizontal *x*-axis. Frequency of occurrence is plotted on the vertical *y*-axis. The frequency is the number of occurrences of the measured variable within a specific class interval (for example, number of hotels with rooms available in a given price category).

Constructing a histogram

As Figure 6.14 illustrates, the method you choose to construct your histogram can have a significant effect on the appearance of the graph you finally produce. Figure 6.14(a), Figure 6.14(b) and Figure 6.14(c) were drawn using the same data (from Table 6.2). However, each was drawn using different methods of calculating class intervals and frequency distributions. Figure 6.14(a) splits the data range evenly on the basis of the *number* of *x*-axis classes desired. Figure 6.14(b) shows the data on the basis of the desired *size* of the *x*-axis classes (in this case $55) and Figure 6.14(c) is the product of *minimising in-class variations* while maximising between-class variations.

Table 6.2 Data set for histogram construction: *Single night, double room hotel accommodation rates (NZ$), Wellington, 1994*

109	253	118	56
112	124	60	45
95	162	90	59
45	198	100	50
136	156	315	50
156	105	144	65
253	118	101	55
105	152	50	80

Source: New Zealand Tourism Board (1995).

The first two methods of working out class intervals and frequency distributions, which allow you to summarise the data to be depicted in your histogram, require that you calculate the range of the data set. **Range** is the difference between the highest data value and the lowest data value. To illustrate, consider the data shown in Table 6.2, which displays the price of hotel accommodation in Wellington, New Zealand during 1994.

Range is the difference between the highest data value and the lowest.

The most expensive room rate in Wellington in 1994 was $315 and the lowest was $45. Therefore, the range is:

$$\$315 - \$45 = \$270$$

The next step is to calculate **class intervals**.

Methods of calculating class intervals

1   One common strategy for calculating class intervals is simply to divide the range by the number of classes you wish to portray. The result will be a number of evenly sized classes. For example, if we use the data in Table 6.2, the range is $315 – $45 = $270. You might have decided that you wish to have a histogram with five classes. Divide $270 by 5 and the result is an interval of $54. Thus, we have intervals of:

Class 1 $45–$98
Class 2 $99–$152
Class 3 $153–$206
Class 4 $207–$260
Class 5 $261–$315

The lowest class begins with the lowest value ($45 in this example). To find the *lower limit* of the *next* class we add $54, which produces a figure of $99. We then add $54 to $99 to produce the lower limit of the next class, $153, and so on. The *upper limit* of each class is found by subtracting 1 unit of the measurement form being used (for example, $1, 1 centimetre, 1 metre, 0.01 gram or 1 tonne) from the lower limit of the class above. The upper limit of the lowest class in the example is therefore $98. Repeat this procedure until the intervals for all classes are calculated. Note that *discrete class intervals* are used (for example, $45–$98, $99–$152) rather than $55–$99, $99–$153). In this way there is no confusion about the class within which any data point is placed (for example, in which class would you put a $99 room charge?).

2 An alternative, but closely related, strategy is the one followed in producing Figure 6.14b. Calculate the data range and then think about the character of the data set to be portrayed. Would it be useful to your audience to read the data on a graph that uses intervals of, for example, $10s or $50s rather than the $13s, $77s, $54s and other odd numbers that might be achieved by the simple division of the range by the number of desired class intervals as described in the preceding strategy? Similarly, would it be useful to begin or end the class intervals at some points other than those fixed by the high and low points of the data set? With these thoughts in mind, the following intervals were chosen for the data in Table 6.2.

There are three common ways of working out class intervals in a data set. Choose the one that summarises the data most usefully.

Class 1 $0–$54
Class 2 $55–$109
Class 3 $110–$164
Class 4 $165–$219
Class 5 $220–$274
Class 6 $275–$330

The graph that resulted is Figure 6.14b. As you can see, class intervals that extended in value beyond the upper and lower limits of the data range were selected. The first class begins at $0 and the class interval is $55, which is useful when one is considering the matter of hotel accommodation costs in New Zealand.

3 Yet another technique of working out class intervals is to minimise in-class variations while simultaneously maximising between-class variations. Look for clusters of data points within the total data set and subdivide the data range using those natural breaks into equal size divisions that best discriminate between clusters. A useful tool in this process is the linear plot. Draw a horizontal line and affix to it a scale sufficient to embrace the maximum and minimum values of the data. Locate each of the data points

on the scale with a short vertical line. If you are using the linear plot for presentation purposes, rather than for calculation only, you should also label each of the data points and provide a title and source. The plot will graphically portray the data distribution (see Figure 6.15).

Figure 6.15   Example of a linear plot: *Accommodation rates (NZ$)*

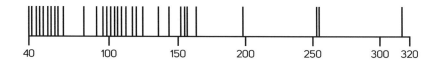

In this example, the data is clustered quite heavily in the range $50–$150. It might be appropriate to produce a histogram that breaks the data into the following ranges, as depicted in Figure 6.14(c):

Class 1 $40–$69
Class 2 $70–$99
Class 3 $100–$129
Class 4 $130–$159
Class 5 $160+

There is no definite rule governing the number of classes in a frequency distribution. Choose too few and information could be lost through a large summarising effect; that is, the picture will be too general. With a lot of classes, too many minor details may be retained, thereby obscuring major features.

Once you have worked out class intervals, the next step in the construction of a histogram is the creation of a frequency table that will allow you to work out the total number of individual items of data that will occur in a particular class.

Constructing a frequency table

As shown in Table 6.3, constructing a frequency table is simply a matter of going through the set of data, placing a tally mark against the class into which each datum falls, then summing the tally to find the frequency with which values occur in each class. If you are working with particularly large data sets, Microsoft Excel has a tool for creating frequency tables that may be helpful. Table 6.3 uses the class intervals described in point 2 of the 'class intervals' discussion above.

Table 6.3 Frequency table of data from Table 6.2

Classes ($)	Tally of occurrence	Frequency
0–54	III	3
55–109	NN NN NN	15
110–164	NN NN	10
165–219	III	3
220–274	II	2
275–330	I	1

The observed frequency is then plotted on the y-axis of the histogram and the classes are plotted on the x-axis (see Figure 6.14b). The rectangles that result should touch each other, thus reflecting the continuous nature of the observations.

Population pyramids or age–sex pyramids

Population pyramids are a form of histogram used to show the number or, more commonly, the percentage of a population in different age groups of the total population. They also illustrate the female–male composition of that population. Figure 6.16 is an example of a population pyramid.

Constructing a population pyramid

By convention, a population pyramid shows males on the left and females on the right.

Although a population pyramid is a form of histogram, a few peculiarities do bear noting. As Figure 6.16 illustrates, a population pyramid is drawn on one vertical axis and two horizontal axes. The vertical axis represents age and is usually subdivided into five-year age cohorts (for example, 0–4, 5–9, 10–14 years). The size of those cohorts may be changed (for example, to 0–9, 10–19 or 0–14, 15–19) depending on the nature of the raw data and the purpose of your pyramid. Remember from the discussion of histograms, however, that if you depict different size cohorts in the same pyramid, the area of each bar must reflect that variability. For example, a 0–14 cohort would be half as wide as a 15–44 cohort on the same graph. On either side of the central vertical axis are the two horizontal axes. That on the left of the pyramid shows the percentage (or number) of males while that on the right shows the figures for females. You will also see from Figure 6.16 that the zero point for each of the horizontal axes is in the centre of the graph. As a final note, it may also be helpful for your reader if you include a statement within the graph of the total population depicted.

Figure 6.16 Example of a population pyramid: *Age-sex structure of Adelaide Statistical Division, 2006*

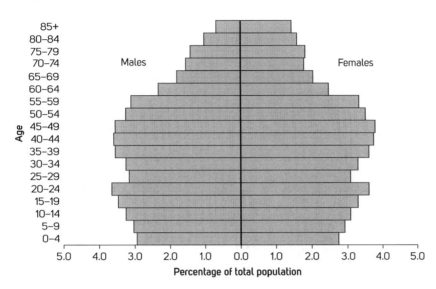

Total population = 1 095 421.

Source: Australian Bureau of Statistics (2006).

Circle or pie charts

Pie charts show how a whole is divided up into parts and what share or percentage belongs to each part. Pie charts are a dramatic way of illustrating the relative sizes of portions of some complete entity (Windschuttle & Windschuttle 1988, pp. 272–3). For example, a pie chart might show how a budget is divided up or who receives what share of some total. See Figure 6.17 for an example.

Constructing a pie chart

Constructing a pie chart usually requires a little arithmetic. It is necessary to match the 360° that make up the circumference of a circle with the percentage size of each of the variables to be graphed. Simply, this is achieved by multiplying the percentage size of each variable by 3.6 to find the number of degrees to which it equates. Obviously, if the values have not already been translated into percentages of the whole, this will need to be done first.

For example, at the 1981 Census there were 1 319 327 people in South Australia and 14 926 786 in the whole of Australia (that is, the South Australian population was 8.8 per cent of the national total). If a pie diagram of the Australian population by states and territories were to be drawn, the segment representing South Australia would have an angle of:

$$\frac{1\,319\,327}{14\,926\,786} \times 360 = 31.8°$$

Figure 6.17   Example of a pie chart: *Primary land use in Murray–Darling Basin, 2000*

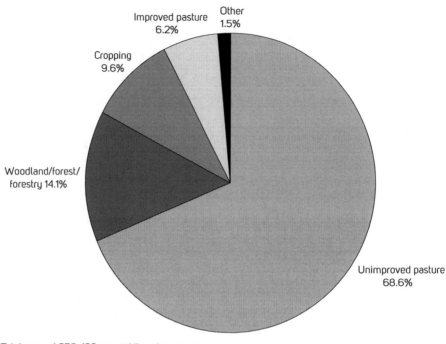

Total area = 1 056 420 square kilometres.

Data source: Atech Group (2000, p. 61).

It is best not to have too many categories (or 'slices') in a pie chart as this creates visual confusion. Five or six segments would seem to be a fair maximum. Generally, no segment should be smaller than 6°. This may require that some classes be grouped together.

The sectors in a pie chart normally run clockwise, with the largest sector occurring first (Australian Bureau of Statistics 1994, p. 117). The starting point for the first sector is created by drawing a vertical line from the centre of the circle to the 12 o'clock position on the circumference.

A pie chart should always make clear the total value of all categories it illustrates.

Pie charts should also advise the reader of the *total value* of categories plotted, as shown, for example, by the statement in Figure 6.17 of the total land area. There is little point in letting a reader know percentages without allowing them the opportunity to determine exactly how much that percentage represents in absolute terms.

Logarithmic graphs

Logarithmic graphs are used primarily when the range of data values to be plotted is too great to depict on a graph with arithmetic axes (commonly a scattergram or line graph). Comparative national Gross Domestic Product figures are good examples of such data, for national figures range from millions of dollars to billions and trillions of dollars. Similarly, historical population figures, which might grow from hundreds to thousands to millions, sometimes necessitate the use of a logarithmic graph. Figures 6.18a and 6.19 are examples of logarithmic graphs.

Figure 6.18   Comparison of data displayed on semi-log and arithmetic graph paper

(a) *South Australia's resident population, 1841–2001*

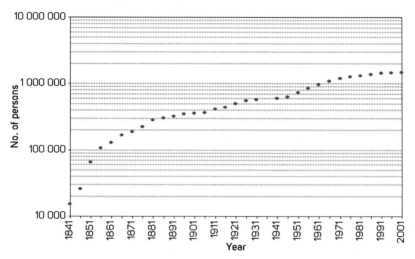

(b) *South Australia's resident population, 1841–2001*

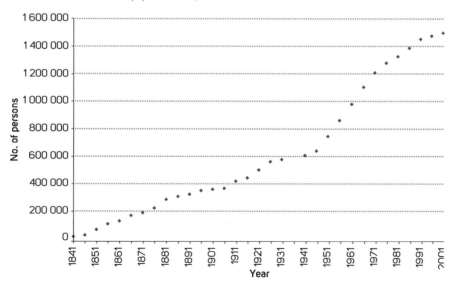

(a) = semi-log graph paper.

(b) = arithmetic graph paper.

Data source: Australian Bureau of Statistics (2001, p. 128).

Logarithmic graphs are good for depicting vast data ranges.

Logarithmic graphs are sometimes also used to compare rates of change within and between data sets. Despite vast differences in numbers, if line *slopes* in a logarithmic graph are the same, then the *rate* of change is similar. This might be useful, for example, if one was illustrating historical rates of population change in a region and trying to argue that despite the fact that the population is now growing at millions of people per year, the *rate* of change has not actually altered since the late 1800s, when the population was growing by thousands each year (see Figures 6.18a and 6.18b, for example).

Before discussing this form of graph any further, it might be helpful to say a little about logarithms. The logarithm of a number (in base 10) is the power to which 10 must be raised to give that number. For example, the log of 100 is 2 because $10^2 = 100$ (that is, 10 raised to the power 2 = 100). Thus, the log of 10 is 1, the log of 100 is 2, the log of 1000 is 3 and so on.

Semi-log graphs have one logarithmic axis; log–log graphs have two.

Second, simple line graphs and scattergrams, as described earlier, typically use an **arithmetic scale** on their axes (for example, 1, 2, 3, 4 … or 0, 2, 4, 6 …) where a constant numerical difference is shown by an equal interval on the graph axes. In contrast, semi-logarithmic and log–logarithmic graphs use a **logarithmic scale** where the numerical value of each key interval on the graph increases *exponentially* (for example, 10, 100, 1000, 10 000 …) and the lines in each cycle (each cycle is an exponent of ten) of the graph become progressively closer together (see Figures 6.18a and 6.19 for examples). Figure 6.18a is an example of a **semi-log graph**. It has a logarithmic y-axis and an arithmetic x-axis. (Histograms and bar graphs can also be drawn using semi-logarithmic paper if the variable to be depicted through the y-axis has a particularly large range.) Figure 6.19 is a **log–log graph**—both axes are logarithmic.

Figure 6.19 Example of a graph with two logarithmic axes (a 'log–log' graph): *Relationship between GDP and passenger cars in use, selected countries, 1992*

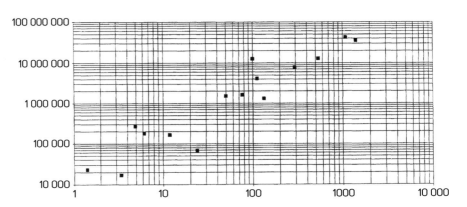

Zero is never used on a logarithmic scale because the logarithm of zero is not defined. Scales on log paper start with ... 0.001, 0.01, 0.1, 1, 10, 100, 1000, 10 000, 100 000 ... (or any other exponent of 10). If you look at Figure 6.19 as an example, you will see that the x-axis commences at 1, whereas the y-axis commences at 10 000. The determination about the figure with which to start the axis is made on the basis of the smallest data point. For example, if the smallest figure for plotting was 35 000, you would start the axis at 10 000, not 1000 or 100 000.

Zero is never used on a logarithmic axis.

Constructing a logarithmic graph

Consider the maximum and minimum values of the data sets to be plotted. In looking, for example, at the population of the state of South Australia over the period 1841–2001, we see that the population grew from 15 000 to 1.5 million people. To accommodate this range of data, the graph will need to have three logarithmic cycles, with the first starting at 10 000, the second at 100 000, and the third at 1 000 000. Because the years are plotted arithmetically, the graph will be drawn in semi-log format. Simply plot data points at their appropriate (x,y) coordinates. Add an appropriate title and indication of data source. The same procedures apply for log–log graphs except, of course, that it is necessary to consider the number of cycles for both axes, not just one.

Preparing a good table

Tables present related facts or observations in an orderly, unified manner. They are used most commonly for summarising results. Tables can be effective for organising and communicating large amounts of information, especially numerical data, although you should not make the mistake of trying to communicate too much information at once.

The main reasons for using tables are:

- to facilitate comparisons
- to reveal relationships
- to save space.

Table 6.4   Example of a table: *Major groundwater resources of Australian states/territories, 1987*

State/ territory	Area of aquifers (km²)	Ground water resource (gigalitres)				
		Fresh	Marginal	Brackish	Saline	Total
New South Wales	595 900	881	564	431	304	2180
Victoria[1]	103 700	469	294	691	30	862
Queensland	1 174 800	1760	683	255	144	2840
South Australia	486 100	102	647	375	86	1210
Western Australia	2 622 000	578[2]	1240	652	261	2740
Tasmania	7240	47	69	8	—	124
Northern Territory	236 700	994	3380	43	10	4420
Australia	**5 226 440**	**4831**	**6877**	**1833**	**835**	**14 376**

[1] In case you had not noticed, a note identifier was inserted at the the entry for Victoria simply to illustrate how a table footnote might look. See the discussion below to find out about the purpose of such notes.
[2] Look, I did it again—this time beside the freshwater data entry for Western Australia.

Source: Australian Bureau of Statistics (1994, p. 18).

A good table should be comprehensible in its own right, requiring no additional explanation to be understood.

Tables should be self-explanatory. This requires a comprehensive title and good labelling for the data in columns and rows. Data shown in the table may be referred to and discussed in accompanying text, but should not be repeated extensively. Table 6.4 is an example of a well laid-out table.

Elements of a table

Aside from the information being conveyed, the main elements of a table are:

1   *Table number*—each table should have a unique number (for example, Table 1) allowing it to be referred to easily in the accompanying text.
2   *Table title*—the title, which is placed one line above the table itself, should be brief and allow any reader to fully comprehend the information presented without reference to other text. The title should answer 'what', 'where' and 'when' questions.

3 *Column headings*—headings are necessary to explain the meaning of data appearing in the columns. It is a good idea to specify the units of measurement (for example $, millimetres, litres) within the column headings (only a small amount of space is available for headings so they must be concise; however, not so concise that they become ambiguous). Below the column headings a dividing line is placed to separate them from the data. The bottom of the table should be marked with a single horizontal line.

4 *Row labels*—like column headings, these explain or define the row data.

5 *Table notes*—notes appear below the table to provide supplementary information to the reader, such as restrictions that apply to some of the reported data. Table notes may also be used for explanation of any unusual abbreviations or symbols.

6 *Table source*—an indication of the source from which the data was derived or the place from which the table was reproduced should be provided. An accepted form of referencing must be used. See Chapter 10 on referencing for more information.

In the end, always be sure that the way you present a table or figure helps your reader understand it completely and correctly. If you have any doubts, present the figure or table to a friend or your lecturer as a stand-alone document and ask them if it makes sense!

REFERENCES AND FURTHER READING

Anderson, J. & Poole, M. 2002, *Assignment and Thesis Writing*, 4th edn, John Wiley, Milton.
Includes a useful and detailed discussion in chapter 12 on preparation and use of tables.

Atech Group 2000, *Aggregated Nutrient Emissions to the Murray–Darling Basin*, Prepared for the National Pollutant Inventory, Environment Australia, Canberra.

Australian Bureau of Statistics 1994, *Australia Yearbook, Cat. no. 1301*, AGPS, Canberra.

Australian Bureau of Statistics 2001, *South Australia: A Statistical Profile 2001*, Cat. no. 1368.4, Australian Bureau of Statistics, Adelaide.

Australian Bureau of Statistics 2005, *Labour Force Australia, Detailed: Electronic delivery*, Cat. no. 6291.0.55.001, Table 08, 'Employed Persons by status in employment and Sex', Australian Bureau of Statistics, Canberra.

Australian Bureau of Statistics 2006, *Census of Population and Housing: Time Series Profile, 2006*, Cat. no. 2003.0, viewed 8 December 2011, <www.abs.gov.au/ausstats/abs@.nsf/productsbytopic/3177338022D899BACA2570D90018BFAE?OpenDocument>.

Australian Bureau of Statistics 2011, *Building Approvals, Australia, August 2011*, Cat. no. 8731.0, Australian Bureau of Statistics, Canberra.

Australian Institute of Health and Welfare 2011, *State and Territory life expectancy*, viewed 8 December 2011, <www.aihw.gov.au/state-and-territory-life-expectancy>.

Bureau of Meteorology 2004, *Climate averages for Darwin airport*, Commonwealth of Australia, viewed 20 April 2005, <www.bom.gov.au/climate/averages/tables/cw_014015.shtml>.

Cartography Specialty Group of the Association of American Geographers 1995, 'Guidelines for effective visuals at professional meetings', *AAG Newsletter*, vol. 50, no. 7 (July), p. 5.

Coggins, R.S. & Hefford, R.K. 1973, *The Practical Geographer*, 2nd edn, Longman, Camberwell, Victoria.

Commonwealth Bureau of Census and Statistics (Australia) 1965, *Overseas arrivals and departures March quarter, 1965*, viewed 8 December 2011, <www.abs.gov.au/AUSSTATS/abs@.nsf/DetailsPage/3401.01965?OpenDocument>.

Eisenberg, A. 1992, *Effective Technical Communication*, 2nd edn, McGraw-Hill, New York.

Forster, C.A. 2004, *Australian Cities: Continuity and Change*, 3rd edn, Oxford University Press, South Melbourne.

Gustavii, B. 2008, *How to Write and Illustrate a Scientific Paper*, 2nd edn, Cambridge University Press, New York.
See chapters 5–8 of this excellent book for advice on preparing figures, diagrams and tables.

Hamblin, A. 1998, *Environmental Indicators for National State of the Environment Reporting—The Land, Australia: State of the Environment (Environment Indicator Reports)*, Department of the Environment, Canberra.

Kosslyn, S.M. 2006, *Graph Design for Eye and Mind*, Oxford University Press, New York.
This book gives attention to the needs and abilities of the audiences of graphs and presents eight psychological principles for constructing effective graphs.

Krohn, J. 1991, 'Why are graphs so central in science?', *Biology and Philosophy*, vol. 6, no. 2, pp. 181–203.
This paper critically questions the prominence, use and significance of graphics in science.

McDonald, J.H. 2009, 'Guide to fairly good graphs with Excel', *Handbook of Biological Statistics*, viewed 1 June 2011, <http://udel.edu/~mcdonald/statgraph.html>.
Helpful online resource offering general tips on graph selection and presentation as well as more detailed advice in graph preparation using Microsoft Excel.

Mohan, T., McGregor, H. & Strano, Z. 1992, *Communicating! Theory and Practice*, Harcourt Brace, Sydney.

Moorhouse, C.E. 1974, *Visual Messages*, Pitman, Carlton, Victoria.
Chapter 7 is a very handy piece on graphics. Well worth considering.

New Zealand Tourism Board 1995, *New Zealand Where to Stay Guide*, New Zealand Tourism Board, Wellington.

Nicol, A.A.M. & Pexman, P.M. 2010, *Displaying Your Findings: A Practical Guide for Creating Figures, Posters, and Presentations*, 6th edn, American Psychological Association.
An extensive and very popular volume that devotes a great deal of attention to the preparation and presentation of high-quality figures and tables (as well as other graphic communications such as posters and photographs).

Pechenik, J.A. 2009, *A Short Guide to Writing About Biology*, 7th edn, Pearson Longman, New York.

Queensland Department of Environment and Resource Management 2011, *Rural water use statistics*, viewed 8 November 2011, <www.derm.qld.gov.au/rwue/statistics.html>.

Rowntree, D. 1990, *Teaching through self-instruction*, Kogan Page, London.

UNESCO 2011, *World Heritage List Statistics*, viewed 8 December 2011, <http://whc.unesco.org/en/list/stat/>.

United Nations 2004, *World Population Prospects: The 2004 Revision, Highlights*, United Nations Department of Economic and Social Affairs, Population Division, New York (online), viewed 20 April 2005, <www.un.org/esa/population/publications/WPP2004/2004Highlights_finalrevised.pdf>.

United Nations Statistics Division 2011, *Demographic Yearbook 2009–2010*, viewed 8 December 2012, <http://unstats.un.org/unsd/demographic/products/dyb/dyb2009-2010.htm>.

Wainer, H. 1984, 'How to display data badly', *The American Statistician*, vol. 38, no. 2, pp. 137–47.
A fascinating review of twelve techniques for displaying data badly! Well worth reading.

Windschuttle, K. & Windschuttle, E. 1988, *Writing, Researching, Communicating*, McGraw-Hill, Sydney.

SEVEN

Making a map

I have an existential map. It has 'You are here' written all over it.

Steven Wright

A map is often the heart, or better, the brain, of a scientific paper.

Paraphrase of Morgan, in Day (1989, p. 56)

Key topics

- Different types of maps
- Standard elements of good maps

Geographers would be lost without **maps**. Put simply, 'A map is a graphic device to show where something is' (Moorhouse 1974, p. 86). In slightly more complex, though more informative, terms a map is 'a scaled, symbolic representation of part of any surface (usually the surface of the Earth) containing selective information and expressed using culturally determined conventions' (Boyd & Taffs 2003, p. 47). Maps use labels, symbols, patterns and colours to convey messages about spatial and other relationships. Maps are marvellous devices for exploring research questions and for pointing out relationships that might otherwise be difficult to see. Indeed, maps can provoke more research questions than they resolve! This chapter outlines different types of maps and offers advice on ways of producing them.

Different types of maps

There are many different types or classes of map (see Robinson et al. 1994 and Slocum et al. 2009 for a discussion). Table 7.1 provides a summary of the nature and function of several common types of map. The rest of this chapter describes more fully the first three of these and offers some advice on their construction.

Table 7.1 Some common types of maps and their nature or function

Type of map	Nature or function
dot	A map that uses dots to illustrate spatial distribution of discrete data by unit of occurrence (e.g. one dot represents one person) or some multiple of those units (e.g. one dot represents 1000 sheep).
choropleth	A shaded or cross-hatched map used to display statistical distributions (e.g. rates, frequencies or ratios) on the basis of areal units such as nations, states and regions.
isoline	A map that shows sets of lines connecting points of known, or estimated, equal values (e.g. elevation, barometric pressure or temperature).
cadastral	A specialised map showing surveyed land tenure; from *cadastre*, an official register or list of property owners and their holdings (Robinson et al. 1994, p. 11).
orthophoto	A map created from a mosaic of aerial photographs and overlain with information such as contours, transport routes and place names. The Land Information New Zealand (LINZ) website at <www.linz.govt.nz> provides some examples of these.
topographic	A common, general-purpose map that typically depicts contours, physical (e.g., rivers and peaks) and cultural features (e.g., roads, churches and cemeteries). The Geoscience Australia website at <www.ga.gov.au/products-services/maps.html> offers online examples of topographic maps for Australia.

Dot maps

Dot maps are a common way of showing both the spatial distribution and the quantity of a variable. Figure 7.1 is an example of a dot map. A dot representing every occurrence of a given characteristic or some multiple of that occurrence is placed on the map. For example, a single dot might represent a single factory, 1000 people or 300 kangaroos. Dots are useful for showing the distribution of discontinuous or discrete data sets (for example, population or stock numbers). It is worth noting that the symbol used need not actually be a dot. Some dot maps substitute other symbols. So, for example, a map of university campus locations might use a mortar board symbol, while a map of poultry farms might use a chicken symbol. Of course, a key or legend will need to be included in any such map to explain what the symbol means.

Figure 7.1 Example of a dot map: *Distribution of Vietnamese-born people, Adelaide Statistical Division, 1996*

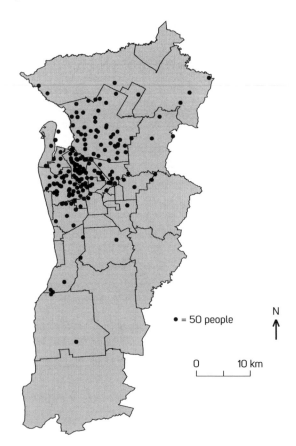

• = 50 people

N

0   10 km

Data source: Australian Bureau of Statistics (1996).

Constructing a dot map

1 Decide on the number of units each dot will represent. The scale of dots needs to be chosen carefully if it is to be effective. If the value of each dot is too large, sparsely populated districts may not be represented at all. If the value is too small, dots will join in densely populated districts (Garnier 1966, p. 16) and you are likely to go crazy drawing the map. The size of the actual dot is also important. If too large, it will make the map look messy and crude. If too small, it will fail to depict spatial variation. Try plotting the greatest and lowest densities to be shown on the map (see the next step for some help on that) using several different dot values to get some impression of the value that will be most effective.

2 Dot maps are generally divided into areas such as statistical divisions or administrative units. For each division, calculate the number of dots to be shown and pencil in those numbers on your map.

3 Draw the appropriate number of dots, ensuring that they are spaced evenly (but not in lines). However, if variations are known to occur within the area, attempt to locate the dots in ways that best represent the spatial variations you know about. For instance, in drawing a dot map of the population distribution of Australia or New Zealand, it would be inappropriate to distribute dots evenly across the entire country. Instead, dots would be placed in focal areas near major population centres.

4 Complete the map by adding a title, legend, scale and source of data.

Proportional circle maps

Proportional circle maps are a variation on the dot map. Instead of using dots of identical size to represent numbers of some phenomenon, the size of each circle (or other symbol) in a proportional circle map is (as the name suggests) directly related to the frequency or magnitude of the phenomenon represented. Figure 7.2 is an example of a proportional circle map.

Constructing a proportional circle map

Circles may illustrate quantities on maps by using a scale that is related to either:

1 the *diameter* of a circle
2 the *area of a circle*.

Both methods are discussed below. The area-based representation of quantities is preferable as it provides a more accurate visual portrayal of a quantity than the diameter-based approach.

Figure 7.2 Example of a proportional circle map: *Distribution of England-born people by Statistical Local Areas for Adelaide and Statistical Subdivision for South Australia, 1996*

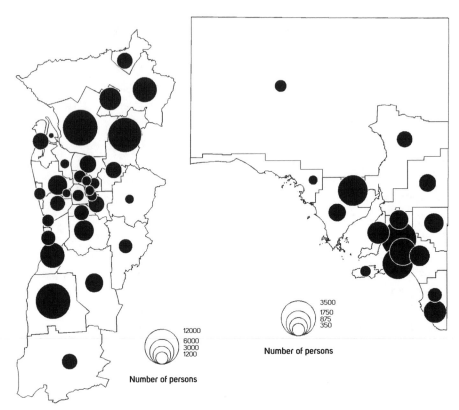

Source: Beer & Cutler (1999, p. 37). Copyright resides with Commonwealth of Australia. Used with permission.

Proportional circles—diameter-based

Area-based proportional circles show quantity more accurately than diameter-based circles.

If the populations of two small towns (for example, Port Gerard with 400 people and Susanville with 1600 people) are plotted using this form of proportional circle, the *diameter* of the circle representing Susanville should be four times as large as that of the circle representing Port Gerard. The diameter of the circle is directly proportional to the given value. Thus, if you have decided that each millimetre of circle radius will represent 100 people, the circle diameter for Port Gerard will be 4 millimetres while that of Susanville will be 16 millimetres (Garnier 1966, p. 18). This is a simpler method of drawing proportional circles than the area-based technique, but it overemphasises the visual impact of large values because circle areas grow exponentially with increases in radius. For this reason, Figure 7.2 does not use this method. Instead, it uses the area-based approach.

Proportional circles—area-based

In this method, statistics are portrayed by circles whose *areas* are proportional to the size of the variables being depicted. This is achieved by making the diameter of the circle proportional to the square root of the number being illustrated. Taking the example discussed above, the area of the larger circle should be four times that of the smaller one. The square root of Port Gerard's population of 400 is 20 and the diameter of that circle will be proportional to that number. The square root of Susanville's population of 1600 is 40 and, following either the graphical or mathematical method of calculating circle size outlined below, an appropriately sized circle will be drawn. Even though the square root of Susanville's population is only twice as large as that of Port Gerard, the circle area will be four times greater thanks to the magic of mathematics. The end result is two circles whose areas give an accurate visual representation of the fact that one town has four times as many people as the other. Figure 7.2 uses proportional circles that are area-based.

To create area-based proportional circles using a graphic method, follow these steps:

1 Decide on the radius of the biggest circle that can be used on your map or figure. This will depend on the scale of the map and the number of proportional circles to be shown. Aim for a result that is a happy medium between a small number of insignificant circles and something that looks like a bubble-bath.
2 Calculate the square root of each of the quantities to be illustrated.
3 Construct a continuous scale of circle size from which you can read off the radius of any of the circles to be drawn. (It may be useful to consult Figure 7.3 as you read this section and the next.) To make this scale, find yourself a spare sheet of paper and create a horizontal arithmetic scale that:
   a is divided into equal units
   b begins at zero
   c embraces the whole range of the square roots you calculated in the first step.
4 In step 1 you decided what the maximum circle radius should be. Now draw a vertical line of that length upwards from your horizontal arithmetic scale at the point corresponding with the largest square root value you calculated in step 2. Join the top of that vertical line to the zero point on the horizontal axis. Believe it or not, the result should be a triangle that will allow you to read off required radii from the horizontal axis without any further calculations. See Figure 7.3 for an example.

Figure 7.3 Example of a graphic scale for creating proportional circles

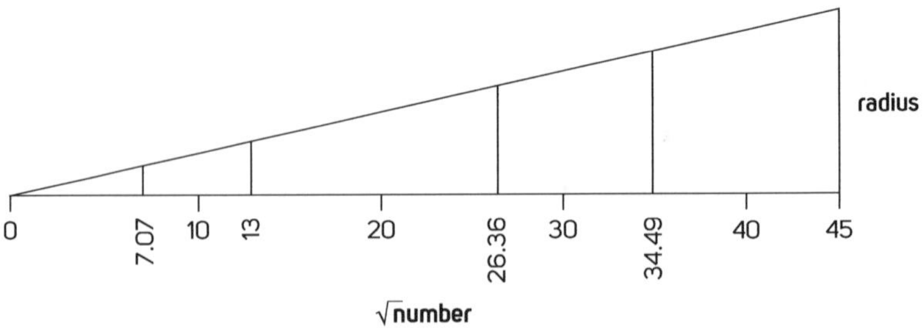

radius

0   7.07   10   13   20   26.36   30   34.49   40   45

√**number**

Include both a circular scale and a linear scale on a proportional circle map.

5 To use the scale you have created, place the point of your compass on that part of the horizontal axis that corresponds to the square root of the value to be represented. Open the other leg of the compass to reach that point of the diagonal line directly above the square root value to be plotted. With the compass set at this radius, draw a circle on the map (or illustration) centred on the location of the area being plotted.

6 A circular scale must be shown on the finished map. A neat and simple way of doing this is to draw circles corresponding to rounded representative figures in the data (not the square root values). See Figure 7.4 for an example.

Do *not* use the triangular scale you created to construct the proportional circles as the scale in your illustration.

Figure 7.4 Example of a completed proportional circle scale

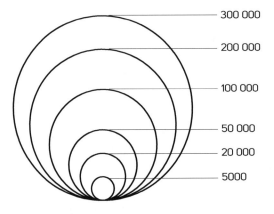

300 000

200 000

100 000

50 000

20 000

5000

Following International Cartographic Association (1984, p. 106) advice, proportional circle areas can also be calculated mathematically. This is a relatively straightforward procedure. Say, for example:

N = maximum value to be represented

n = one of the other values in the data series

S = the area of the circle representing N
s = the area of the circle representing n
R = the radius of the circle representing N
r = the radius of the circle representing n

$$\text{then} \quad \frac{s}{S} = \frac{n}{N} \quad \text{or} \quad \frac{\pi r^2}{\pi R^2} = \frac{n}{N} \quad \text{thus} \quad r = \frac{R\sqrt{n}}{\sqrt{N}}$$

You will see that this procedure still requires you to follow steps 1 and 2 outlined in the graphic method of calculation. Of course, you will also need to prepare a scale to put on the completed map (step 6 above).

The technique of using proportional circles may be usefully extended to represent two quantities (for example, freight tonnages in and out of a region) in the form of a *split proportional circle*. To achieve this, follow the instructions above, but draw two semicircles at each point. Figure 7.5 provides an example.

Figure 7.5   Example of a split proportional circle

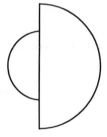

The semicircle on one side might, for example, represent freight into a region or births in a place, whereas the semicircle on the other side could represent freight exports or deaths. The circles are made proportional to the figures being illustrated in exactly the same way as discussed above, but only half of the circle is actually drawn. The same scale must be used for the quantities in each half or comparison will be impossible.

Another way of displaying information on maps is as a *proportional pie diagram*, where the size of the circle shows the magnitude of a given value. However, the circle is then subdivided into different sized segments, each of which represents some percentage of the total value (see the discussion on pie charts in Chapter 6 for more information). For example, a map of housing availability in Darwin might show proportional circles indicating the number of homes in each suburb; these, in turn, might be internally subdivided to show the proportions of government housing, owner-occupied housing, rental dwellings and so on in each of the mapped suburbs.

Proportional pie charts are a useful way of showing magnitudes and proportions simultaneously.

Choropleth maps

A **choropleth map** displays spatial distributions by means of cross-hatching, intensity of shading or colours. Figure 7.6 is an example. Choropleth maps are commonly used to display rates, frequencies and ratios, such as marriage and divorce rates, population densities, birth and death rates, and percentages of total population classified according to sex, age, ethnicity and per capita income. Choropleth maps also may be used to show categorical data such as soil types and vegetation types.

Figure 7.6   Example of a choropleth map: *Unemployment rate, Melbourne Statistical Division, 1996*

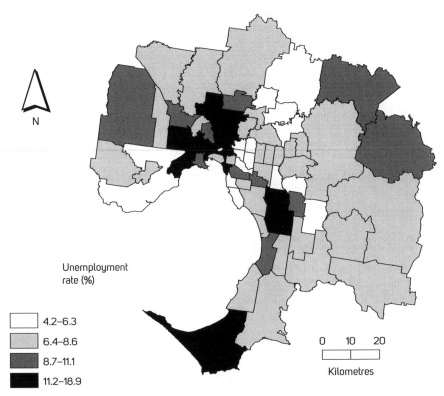

Data source: Australian Bureau of Statistics (1996).

Constructing a choropleth map

The description that follows outlines the preparation of a choropleth map depicting statistical information. Note that maps depicting categorical data require completion of only the last three steps.

Preparing a choropleth map involves a number of basic steps (Sullivan 1993, pp. 69–71).

1   Calculate the range of the data to be mapped. Remember, the range is the difference between the highest and lowest value. For example, if the highest value is 250 millimetres and the lowest is 72 millimetres, then the range is 178 millimetres (that is, 250 − 72 = 178).

2   Decide on the number of classes into which the data will be grouped. This will depend on the purpose of the map and the nature of the data. If there are too many classes, the values for specific areas may be difficult to identify. Four to six classes is generally sufficient to identify spatial patterns without making the map too detailed.

Try to use four to six data classes on a choropleth map.

3   Determine the interval or range of values within a single class. To do this, follow the directions outlined in the discussions on class intervals and frequency distributions under 'Constructing a histogram' in Chapter 6. In general, the outcome should be one in which there are approximately equal size classes (in terms of class interval or number of observations in each class) and there should be no vacant classes (Toyne & Newby 1971, p. 86).

4   Create a cartographic pattern (cross-hatching, grey tones or colour) to represent each of the class intervals you have selected. This will be your *legend*. Low values are typically represented by light colours or shading, and high values by darker colours or shading. An easy method is to use a single colour changing progressively from light to dark. Use colours that match the phenomenon being mapped; for example, greens for vegetation or browns for soils. See a good atlas for examples.

5   Transfer the cartographic patterns of the legend to the map. For every geographic unit such as a nation, state, region or shire, colour or shade that unit according to the pattern in the legend. Avoid having blank areas that do not provide any information.

6   Complete the map by adding a title, scale, northpoint etc (see the later section, 'Standard elements of good maps').

Isoline maps

Most of us see a form of **isoline map** every time we look at newspaper and television weather reports. In common with other forms of isoline map, weather maps show sets of lines (isolines) connecting points known, or estimated, to have equal value. Weather maps usually portray atmospheric pressure, but other forms of isoline maps include **topographic maps**, which depict contours of equal elevation, and rainfall maps (see Figure 7.7).

Figure 7.7   Example of an isoline map: *Average annual rainfall, South Australia*

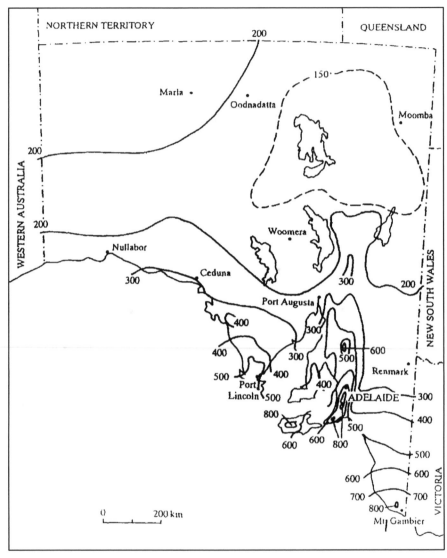

Note: isohyets are in millimetres.

Source: Australian Bureau of Statistics (1994, p. 5). Copyright in ABS data resides with the Commonwealth of Australia. Used with permission.

Isoline maps depict continuous data.   Isoline maps always use data with continuous distributions (for example, rainfall or temperature). Table 7.2 lists common types of isolines and the variables they depict.

Table 7.2 Some common isolines and the variables they depict

Isoline	Connecting places of equal ...
isobar	atmospheric pressure
isotherm	temperature
isohyet	rainfall
contour line	elevation
isobath	water depth

Constructing an isoline map

The following steps are required to prepare an isoline map (based on Garnier 1966).

1 On a base map of the area being represented, locate all points for which precise figures for the phenomenon being mapped are available; see Figure 7.8(a).

2 Calculate the range of the values being mapped and, taking that figure into consideration, decide on a suitable value for the interval between each isoline (that is, for a contour map, should it be 5 metres, 10 metres or 100 metres?). In making this decision, it is helpful to consider the number of known observations upon which you are basing the map. Toyne and Newby (1971, p. 99) suggest that the number of isolines (classes) be no more than five times the logarithm of the number of observations. Thus, for example, if your map has 100 observed points on it, one might expect a map with ten isoline classes. A map with twenty-three known points would have about seven isoline classes. So, if you are drawing a topographic map on the basis of 100 observed points with a data range of 900 metres (say from 800 metres to 1700 metres), it would seem appropriate to aim for an interval between each isoline of about 100 metres (for example, 800 metres, 900 metres, 1000 metres ...). Remember, however, that the interval you choose should be set at a value that will show the detail of the distribution without overcrowding the map. It may be helpful to do several trial plots of the data to determine the best interval to use.

3 Draw the isolines. This is the most difficult part. Disappointing as it may be, the process is not a case of just joining dots. Most point data will not correspond exactly to the contour values decided in the preceding step. The position of isolines between point data of different values is worked out by the technique of interpolation. Figure 7.8(b) shows how this is done for a map of precipitation in South Australia. Assume that the rate of change

Figure 7.8   Creating an isoline map

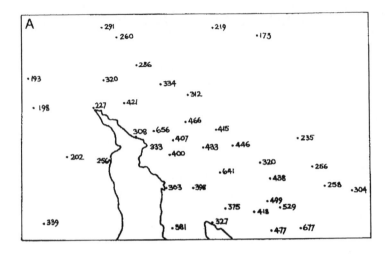

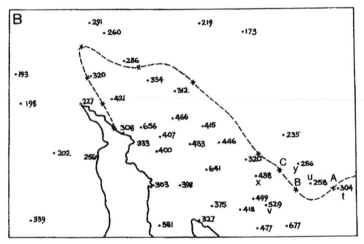

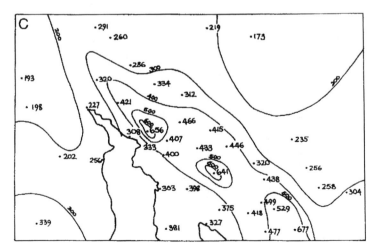

between one known point and another is constant, unless information you have suggests otherwise. Point A, which depicts a isohyet value of 300, is between point u (258 millimetres) and point t (304 millimetres); point B lies about one-quarter of the way between points u (258 millimetres) and v (529 millimetres); and point C is somewhat closer to point y (256 millimetres) than it is to point x (438 millimetres).

4 The appropriate points are then joined by smooth lines as shown in Figure 7.8(c). Make sure that your map has enough isolines to show the data pattern accurately without creating visual confusion. If you have too many isolines, or if the data pattern is not depicted as well as you would like it to be, rethink your selection of isoline interval and redraft the map.

5 Label enough of the isolines with their values to enable readers to understand the map quickly. Place the isoline values so they can be read without turning the map.

6 Complete the map by adding the standard elements (see the following discussion).

Standard elements of good maps

Unless there are good reasons to the contrary, maps should include the following six critical components:

- *map number*—each map should have a unique number (for example, Figure 4), allowing it to be identified in textual discussion.

- *title*—this should be brief yet comprehensible; that is, any reader should be able to understand the information presented without reference to other texts. The title, which is placed above the map itself, should answer 'what', 'where' and 'when' questions.

- *northpoint*—this is an arrow that indicates north, allowing the reader to orient the map correctly. If possible, maps should also make clear the relationship between the area depicted and surrounding territories.

- *scale*—this is an indication of the relationship between distance on the map and distance on the ground. Scale is expressed in one of three ways. It may be expressed as a written statement (for example, '1 centimetre represents 100 kilometres'). It may be depicted graphically, as in Figure 7.1, where a short horizontal line provides an indication of a 10-kilometre distance. Alternatively, it may be expressed as a **representative fraction** (RF), such as 1:50 000 or 1:250 000. An RF of 1:50 000 means that any single unit of measurement—such as 1 centimetre, 1 metre or 1 inch—on the map represents 50 000 of those same units in reality. So, if you are using a 1:50 000 scale map and find you have to hike a mapped distance

A good map should be comprehensible in its own right, requiring no additional explanation to be understood.

of 40 centimetres, how far is the walk in reality? Multiply 50 000 by 40 to find out the distance in centimetres and then convert the answer to a more comprehensible unit of measurement (such as kilometres). Thus, 50 000 × 40 centimetres = 2 000 000 centimetres. This is 20 000 metres or 20 kilometres.

- *legend*—also known as the key, this is an explanation of the symbols, patterns and colours used in a map or diagram.
- *data source*—this is an indication of the source of the mapped data. If you have copied the map from somewhere else, an accepted form of referencing must be used (see Chapter 10 for more information on these two matters). In general, the acknowledgment might take the form of a 'Source: ...' statement on the map. However, if you are producing an original map, perhaps using a Geographical Information System (GIS), you may need to include an extensive array of information about the map, the data and the sources on which you drew to produce it. Typically, the map should have the following details set out on it:
  - Produced by:
  - Data sources:
  - Projection:
  - Spheroid:
  - Datum:
  - Grid:
  - Completed:
  - Project:

Examples of the full array of information expected are set out in Table 7.3. The material on the left-hand side of the box is a complete elaboration of all the details that might be expected. The material on the right shows the ways in which these details are typically presented. As you will see, it is customary to abbreviate more commonly understood terms, such as Universal Transverse Mercator and Geodetic Datum of Australia.

Table 7.3   An example of map production and data details

Details		Typically presented	
Produced by:	A Smith-Brown	Produced by:	A Smith-Brown
Data sources:	Cadastre from DEHAA, SA, 2005	Data sources:	Cadastre from DEHAA, SA, 2005
	Roads from Transport SA, 2004		Roads from Transport SA, 2004

Details		Typically presented	
	Native vegetation from Planning SA, 2005		Native vegetation from Planning SA, 2005
Projection:	Universal Transverse Mercator (UTM)	Projection:	UTM
Spheroid:	Geodetic Reference System (GRS) 1980	Spheroid:	GRS1980
Datum:	Geodetic Datum of Australia (GDA94)	Datum:	GDA94
Grid:	Map Grid of Australia (MGA), Zone 54	Grid:	MGA, Zone 54
Completed:	3 June 2006	Completed:	3 June 2006
Project:	GEOG 3005, 'Introduction to GIS', Exercise 4, Vegetation Mapping	Project:	GEOG 3005, 'Introduction to GIS', Exercise 4, Vegetation Mapping

Finally, it is worth noting that maps should, as far as possible, be sufficiently large to be easily read and interpreted, and that they should be readable from a single viewpoint; that is, a reader should be able to examine the text without having to turn the page sideways. Having said this, labels should also be aligned along linear features (for example, roads), extend over areal features, such as seas and suburbs, and should consistently locate point features (for example, all point features labels might be placed at the four o'clock position relative to the item being labelled)

Make sure your map can be read easily.

REFERENCES AND FURTHER READING

Australian Bureau of Statistics 1994, *South Australian Yearbook*, Cat. no. 1301.4, ABS, Canberra.

Australian Bureau of Statistics 1996, *Census*, viewed 16 January 2012, <www.abs.gov.au>.

Beer, A. & Cutler, C. 1999, *Atlas of the Australian People. 1996 Census. South Australia*, Department of Immigration and Multicultural Affairs, Canberra.

Boyd, B. & Taffs, K. 2003, *Mapping the Environment: A Professional Development Manual*, Southern Cross University Press, Lismore.

Day, R.A. 1989, *How to Write and Publish a Scientific Paper*, Cambridge University Press, Cambridge.

Garnier, B.J. 1966, *Practical Work in Geography*, Edward Arnold, London.

International Cartographic Association 1984, *Basic Cartography for Students and Technicians*, vol. 1, International Cartographic Association, Great Britain.

Map and Chart-Making no date., viewed 3 June 2011, <www.ruf.rice.edu/~feegi/ carto html>
A helpful portal to a range of cartographic resources available on the internet.

Kimerling, A.J., Buckley, A.R., Muehrcke, P.C. & Muehrcke, J.O. 2009, *Map Use, Reading and Analysis*, 6th edn, ESRI Press, Redlands, California.
This marvellous, comprehensive book covers a wide range of material on maps and their uses. A 'must read'.

Krygier, J. & Wood, 2005, *Making Maps: A Visual Guide to Map Design for GIS*, Guilford Press, New York.
Though aimed at producing maps using GIS, this valuable book contains a great deal of helpful advice on key principles in map design including, for example, layout and symbolisation.

Monmonier, M. 1996, *How to Lie With Maps*, 2nd edn, University of Chicago Press, Chicago.
As the title suggests, this is a fascinating and informative book. By focusing on how to lie with maps, it makes good mapping practice clear.

Moorhouse, C.E. 1974, *Visual Messages*, Pitman, Carlton, Victoria.

Robinson, A.H., Morrison, J.L., Muehrcke, P.C., Guptill, S.C. & Kimerling, A.J. 1994, *Elements of Cartography*, 6th edn, John Wiley, New York.

Slocum, T.A., McMaster, R.B., Kessler, F.C. & Howard, H.H. 2009, *Thematic Cartography and Geovisualization*, 3rd edn, Pearson Education, Upper Saddle River, New Jersey.
A high-quality volume that discusses map-meaning and map-making, offering comprehensive advice on traditional and newer forms of map making. A very useful reference.

Sullivan, M.E. 1993, 'Choropleth mapping in secondary geography: An application for the study of middle America', *Journal of Geography*, vol. 92, no. 2, pp. 69–74.

Toyne, P. & Newby, P.T. 1971, *Techniques in Human Geography*, Macmillan, London.

EIGHT

Preparing and delivering a talk

All the great speakers were bad speakers at first.

Emerson in Mohan, McGregor and Strano (1992, p. 331)

An orator is someone who says what he thinks and feels what he says.

William Jennings Bryan

It is not okay to be boring.

Key topics

- Why are public-speaking skills important?
- Preparing to give a talk
- How to deliver a successful talk
- Coping with questions from your audience

Although talking comes naturally to most of us, public speaking remains one of the most frightening things many people can imagine. And, believe it or not, lecturers do appreciate the fact that public speaking is intimidating for many students. While some lecturers have grown accustomed to speaking before a large audience, others still feel some trepidation about speaking to an unknown class, a professional gathering, a community meeting or even a wedding party. They do understand the sleepless nights, sweaty palms, pounding heart, cotton mouth and weak legs that sometimes precede a talk. So, when you are asked by your lecturer to give a prepared talk in class, it is unlikely that the assignment has been selected lightly. Lecturers usually have three fundamental objectives in mind when they ask you to give a talk in your geography or environmental science class.

Why are public-speaking skills important?

First, preparing for and delivering a talk encourages you to organise your ideas, to construct logical arguments and to otherwise fulfil the objectives of a university education.

Public speaking serves valuable intellectual, vocational and professional functions.

Second, your lecturers also have your career interests in mind. Many of the jobs in which university-educated geographers, environmental managers and social scientists find themselves require them to make public presentations. While business and educational leaders acknowledge oral communication and public-speaking skills to be among the most important abilities a university graduate can have, international surveys indicate repeatedly that these skills are also among the most poorly developed. Consequently, your future employers are likely to be impressed if you can point out to them that, through your degree, you have given several multimedia talks to various audiences, particularly if you can explain that you have used each of those opportunities to refine your presentation skills.

Third, developing the ability to speak effectively at conferences and other professional or community gatherings is a critical component of effective geographical or environmental science practice. If we are to make contributions to our significant work in areas such as environmental justice, global warming and indigenous peoples' land rights, it is vital that we be able to share the results of our work in such forums (Hay, Dunn & Street 2005, p. 159). Given the importance of these skills, it would appear that, despite any fears of speaking you might have, you will probably not be spared from having to give one or several 'public talks' of, say, 10 to 20 minutes' duration throughout your degree. The 'mechanics' of giving such a talk are outlined in this chapter, which is divided into three main parts. The first deals with the essentials of preparing for the talk, the second with delivering the talk and the third with coping with post-talk questions. Throughout the chapter, reference is made to the use of Microsoft **PowerPoint** software. This widely used aid to spoken presentations may help you to produce a more effective presentation, as well as handouts for your talk.

The discussion that follows is not intended to be a prescription for a perfect talk. Instead, it offers guidelines to help you prepare for and deliver your first few 'speeches'. With experience you will develop your own style—a form of presentation that may be very effective and yet may transgress some of the guidelines discussed here. Practice will help you to develop your own approach, but you may also want to keep a critical eye on your lecturers and other public speakers that you come across. Pay attention to the form and manner of their delivery. Try to identify those devices, techniques and mannerisms that you believe add to, or detract from, a presentation. Then apply what you learn to your own talks. You might also find it helpful to look over Figure 8.1 to get some idea of the sort of criteria that are important for giving a successful talk.

Figure 8.1　Assessment schedule for a talk

Student Name:	Grade:	Assessed by:

The following is an itemised rating scale of various aspects of a formal talk. Sections left blank are not relevant to the talk assessed. Some aspects are more important than others, so there is no formula connecting the scatter of ticks with the final percentage for the talk. A tick in the left-hand box means that the criterion has been met satisfactorily. A tick in the right-hand box means it has not. If you have any questions about the individual scales, comment, final grade or other aspects of this assignment, please see the assessor indicated above.

	✓	X
First impressions		
Speaker appeared confident and purposeful before starting to speak	☐	☐
Speaker's personal grooming and dress standards of high quality	☐	☐
Speaker attracted audience's attention from the outset	☐	☐
Presentation Structure		
Introduction		
Title/topic made clear	☐	☐
Purpose of the presentation made clear	☐	☐
Organisational framework made known to audience	☐	☐
Unusual terms defined adequately	☐	☐
Body of presentation		
Main points stated clearly	☐	☐
Sufficient information and detail provided	☐	☐
Appropriate and adequate use of examples/anecdotes	☐	☐
Discussion flowed logically	☐	☐

Figure 8.1 *(cont.)*

	✓	✗
Conclusion		
Ending of presentation signalled adequately	☐	☐
Main points summarised adequately/ideas brought to fruition	☐	☐
Final message clear and easy to remember	☐	☐
Coping with questions		
Whole audience searched for questions	☐	☐
Questions addressed in order	☐	☐
Questions handled adeptly	☐	☐
Full audience addressed with answers	☐	☐
Speaker maintained control of discussion	☐	☐
Delivery		
Speech clear and audible to entire audience	☐	☐
Talk given with engagement and enthusiasm	☐	☐
Presentation directed to all parts of audience	☐	☐
Eye contact with audience throughout presentation	☐	☐
Speaker kept to time limit	☐	☐
Good use of time without rushing at end	☐	☐
Pace neither too fast nor too slow	☐	☐
Visual aids and handouts—if appropriate		
Visual aids well prepared	☐	☐
Visual aids visible to entire audience	☐	☐
Speaker familiar with own visual aids (for example, OHTs, PowerPoint images, blackboard diagrams)	☐	☐
Effective use made of handouts and visual aids	☐	☐
Handouts well prepared and useful	☐	☐
Assessor's comments: was this an effective talk?		

Preparing to give a talk

You cannot expect to talk competently 'off the cuff' on any but the most familiar topics. Effective preparation is critical to any successful presentation. Preparation for a talk should begin some days (at least) in advance of the actual event and certainly not just the night before. As part of your preparation, you should:

- establish the context and goals
- organise the material for presentation
- structure your talk
- prepare your text and aids to delivery
- rehearse
- consider final points of preparation.

Establishing the context and goals

- *Who is your audience?* Target the presentation to the audience's characteristics, needs and abilities. The ways in which a topic might be developed will be critically influenced by the background and expertise of your listeners (Eisenberg 1992, p. 333). Find out how big the audience will be, as this may affect the style of presentation. For example, it may be more difficult to lead a successful interactive presentation with a large audience than with a smaller group.
- *Why are you speaking?* The style of presentation may differ depending on your purpose. The purpose may be to present information; to stimulate discussion; to present a solution to a problem; or, perhaps, to persuade a group of the value of a particular view or course of action. Depending on the purpose of the talk, you may have to alter the style and content of your presentation.
- *What is your subject?* Be sure that your subject matches the reason for the presentation. A mismatch may upset, bore or alienate your audience. A clear sense of purpose will also allow you to focus your talk more clearly.
- *How long will you speak?* Confirm how much of the time available is for the talk and how much is intended for audience questions. Avoid the embarrassment of being asked to conclude the talk before it is finished, or of finishing well short of the deadline.
- *Who else is speaking?* This may influence the audience's reaction to you (Eisenberg 1992, p. 332). It may also require that you take steps to avoid repeating things someone else might say.

Try to develop your own style of public speaking, but keep in mind the expectations of your audiences.

To give a successful talk, make sure you know why you are speaking, where, to whom and for how long. Above all, work out what your central message is.

- *Where are you speaking?* Generally, student presentations are delivered in the usual classroom, so you will probably be familiar with the layout of that place. Do take some time, however, to familiarise yourself with audiovisual aids, the location and operation of lighting, and other elements of the room that you may not be accustomed to dealing with. If your talk is to be given in a place other than your classroom, try to visit the venue. Room and layout characteristics can have an effect on the formality of the presentation, the speed of the talk, attentiveness of the audience and the types of audio-visual aids that can be employed. For example, is the talk is to be given in a large room, from a lectern, with a microphone to an audience seated in rows? Box 8.1 demonstrates the need for such reconnaissance by even experienced academics at major professional gatherings!

| Box 8.1 |
| Example of recon-naissance before a talk |

At some point in your career you will likely arrive to present a paper in a theatre or other space that is clearly unsatisfactory, and you will definitely experience technical failures that are beyond your control. In 2003, the Royal Geographical Society and Institute of British Geographers decided to hold the annual geography conference in the Society's august premises located next to the Albert Hall in London. I was to present a paper in a theme session of what was overtly intended as an international conference. Emails from conference organisers had indicated that my paper was to be delivered in Basement Room 1. After a 22-hour flight and a taxi ride, I arrived at the statue of Livingstone outside the Society's premises. It was clear that major renovations were underway, and most of the 'presentation spaces' had been very recently emptied and painted. To find my presentation space, I first went downstairs then, curiously, up to the left via a small stairwell with a distinct Hogwart's feel to it. Basement Room 1 itself was a horror: a claustrophobic broom cupboard without a single screen or piece of technology. When I presented the following morning, the fifteen tightly packed chairs were unable to accommodate delegate demand, while the sad-looking overhead projector that had appeared was permanently and excessively out of focus. I handed out copies of the written paper, which fortunately had the tables and graphics as an attachment, and I presented my paper. Other presenters in similar circumstances were not so lucky as to have hard copy—the ultimate back-up for a conference presentation.

Adapted from: Hay, I., Dunn, K. & Street, A. 2005, 'Making the most of your conference journey', *Journal of Geography in Higher Education*, vol. 29, no. 1, pp. 159–71. Reprinted by permission of the publisher Taylor & Francis Ltd. The full article is available at <www.tandfonline.com>.

Researching and organising the material for presentation

- *Do your research.* Keeping in mind the purpose of your talk, gather and interpret appropriate and accurate information. Make a point of collecting **anecdotes**, cartoons or up-to-date statistics that might make your presentation more appealing, colourful and convincing.
- *Choose the right organisational framework.* Ensure that the framework used is appropriate (see Box 8.2) and that the organisational framework helps to make the point of your presentation clear. For example, if your main aim is to discuss potential solutions to male homelessness in Dunedin, it would probably be less useful to spend most of the time discussing the historical development of social security policies in New Zealand as a prelude to that.

Most presentations seem to adopt one of the following five organisational frameworks:

- chronological—for example, the history of geographic thought from the nineteenth century
- scale—for example, an overview of national responses to desertification followed by detailed examination of responses in a particular area
- spatial—for example, a description of Japan's trading relations with other countries of the Pacific
- causal—for example, the implications of financial deregulation for the New Zealand insurance market
- order of importance—for example, a ranked list of solutions to the problem of male homelessness in Perth.

Box 8.2 Some organisational frameworks for a talk

- *Eliminate superfluous material.* If you have already written a paper upon which your presentation is to be based, be aware that you will not be able to communicate everything you have written. Carefully select the main points and devote attention to the strategies by which those points can be communicated as clearly and effectively as possible. Courtenay (1992, p. 220) makes the following suggestions.
  - List all the things you know or have found out about your subject.
  - Eliminate all those items you think the audience might already know about.
  - Eliminate anything that is not important for your audience to know. Keep doing this until you are left with one or two new and dynamic

points. These should not already be known to your audience and they should be interesting and useful to them. These points should form the basis of your presentation. Similarly, Stettner (1992, p. 226) makes a very good argument for organising a talk around just three main points: no fewer and no more.

- *Give your talk a clear and relevant title.* An audience will be attracted to, and informed by, a good title. Be sure that your title clearly illustrates the subject matter of the talk.

Structuring your talk

A great talk will have a structure that:

- starts with a clear, memorable statement
- concentrates on a small number of key points
- is concise, with an even balance of material from one point to another where balance is appropriate
- is 'signposted'; that is, it highlights connections between the material presented and the talk's main argument or trajectory
- ends with a clear, memorable statement that is consistent with the start (Knight & Parsons 2003, p. 164).

Let us look at the structure of a great talk in more detail. In most cases a talk will have an introduction, a discussion and a conclusion. The introductory and concluding sections of oral presentations are very important. About 25 per cent of your presentation ought to be devoted to the 'beginning' and 'end' (combined). The remaining time should be spent on the discussion.

Introduction

- *Make clear your rationale for speaking and your conceptual framework.* This gives the audience a basis for understanding the ideas that follow. In short, let listeners know what you are going to tell them. Box 8.3 explains how to do this effectively.

**Box 8.3
Introducing
a talk
effectively**

- *State the topic.* 'Today I am going to talk about ...' Do this in a way that will attract the audience's attention.
- *State the aims or purpose.* Why is this talk being given? Why have you chosen this topic? For what reasons should the audience listen?
- *Outline the scope of the talk.* Let the audience know something about the spatial, temporal and intellectual boundaries of the presentation. For example, are you

discussing Australian attitudes to the environment from a Maori perspective, or offering a geographer's view of British financial services in the twenty-first century? Some people love to point out gaps in your work's coverage, so head them off by making clear what you are and are not covering.

- *Provide a plan of the discussion.* Let the audience know the steps through which you will lead them in your presentation and the relationship of each step to the others. It can be useful to show the audience an outline of your talk's structure and progression, perhaps using a PowerPoint slide to do so.

- *Capture the audience's attention from the outset.* Do this with a rhetorical question, relevant and interesting quotes, amazing facts, an anecdote or startling statements. However, avoid jokes unless you have a real gift for humour.

- *Make the introduction clear and lively.* First impressions are very important.

Try to win your audience's attention from the very start of your talk.

Body of the talk

- *Construct a convincing argument supported with examples.* Remember, you are trying to present as compelling a case in support of your findings as possible.

- *Ensure there is a 'fluid logic between your main points'* (Montgomery 2003, p. 172).

- *Limit discussion to a few main points.* It is important to give your audience enough time to absorb the information you present. Lindsay (1984, p. 48) observes that a rule of broadcasting is that it takes about three minutes to put across each new idea. Do not make the mistake of trying to cover too much material. Remember, many of the ideas you discuss may be entirely new to the audience. It is usually better to make a few points clearly, intelligibly and with depth of evidence than it is to deal superficially and hastily with a large array of material.

Limit your talk to discussion of key points and make clear the relationship of each to the overall trajectory.

- *Present your argument logically, precisely and in an orderly fashion.* Try producing a small diagram that summarises the main points you wish to discuss, then use this as a basis for constructing your talk. It might also create a useful handout or PowerPoint slide for your audience.

- *Accompany points of argument with carefully chosen, colourful and correct examples and analogies.* It is helpful to use examples built upon the experience of the audience at which they are directed. Analogies and examples clarify unfamiliar ideas and bring your argument to life.

- *Connect the points of your discussion with the overall direction of the talk.* Remind the audience of the trajectory you are following by relating the points you make to the overall framework you outlined in the introduction. For example, 'Now, moving on to the last of the three points I have identified as explaining ...'

- *Restate important points.*
- *Personalise the presentation.* This can add authenticity, impact and humour. For example, in discussing problems associated with administering a household questionnaire survey, you might recount an experience of being chased down dark suburban streets by a large, ferocious dog. Avoid overstepping the line between personalising and appearing self-centred by ensuring that your tales will help the audience to understand your message.

Conclusion

- *Signal the conclusion.* Phrases like 'To conclude ...' or 'In summary ...' have a remarkable capacity to stimulate audience attention. When you tell your audience you are about to finish, you get their attention!
- *Bring ideas to fruition.* Let the audience know exactly what the take-home message is. Restate the main points in words other than those used earlier in the discussion, develop some conclusions and review implications. Connect your talk with its wider context.
- *Tie the conclusion neatly together with the introduction.* The introduction should have noted where the talk would be going. The conclusion should remind the audience of the content and dramatically observe your arrival at the foreshadowed destination.
- *Make the conclusion emphatic.* Do not end with a whimper! A good conclusion is very important to an effective presentation, reinforcing the main idea or motivating the audience (Eisenberg 1992, p. 340). Use the conclusion to reinforce your main ideas or to motivate the audience. For instance, if you have been stressing the need for community involvement in reducing greenhouse gas emissions, try to 'fire up' the members of the audience so that they feel motivated to take some action of their own.
- *End the presentation clearly.* Unless you make it clear to your audience that your talk is over, they will not know whether to clap (one hopes) or to await more of your talk. Try to avoid giggling self-consciously and muttering things like 'Well, that's the end'. Instead, a phrase like 'Thank you' makes it clear to the audience that your talk is over and is likely to encourage applause that recognises the effort you have put into your presentation.

Preparing your text and aids to delivery

Different people have different preferences when it comes to preparing a presentation 'script'. Some opt for a full script, others for brief notes, and still others for graphic images such as a flow diagram. The same model is not appropriate for everyone, so heed Montgomery's (2003, p. 171) advice to 'design and write out your talk in a manner you feel comfortable with'. That said, there are a few points worth considering.

Preparing notes

- *Prepare well in advance.* Mark Twain is reported to have said 'It usually takes more than three weeks to prepare a good impromptu speech' (Windschuttle & Elliott 1994, p. 341). Twain may have overstated the case a little, but it is fair to consider the talk as the tip of the iceberg and the preparation the much larger submerged section.
- *Prepare a talk, not a speech.* In general, you should avoid preparing a full text to be read aloud. A presentation that is simply read aloud is often boring and lifeless, with the speaker better connected with the script than with the audience. If you must prepare a text to be read, remember that a talk needs to be kept simple and logical. Because your talk will go past your listener only once, it must also 'be very well organized, developed logically, stripped of details that divert the listener's attention from the essential points of the presentation' (Pechenik 2004, p. 252).
- *Write for speaking, not reading.* Most people speak at between 125 and 175 words per minute (Dixon 2004, p. 100), though a good speed for formal presentation delivery is about 100 words per minute (Montgomery 2003, p. 171). So, if you feel absolutely compelled to write a script, you will know that a 10-minute talk will require you to prepare about 1000 words. Keep sentences short and simple. Major points need to be restated. Language should be informal, but should not employ slang and other conventions of cafe conversation and bar-room banter.
- *Prepare personal memory prompts.* These might take the form of clearly legible notes, key words, phrases or diagrams to serve as the summary outline of your talk. Put prompts on cards or on notepaper, ensuring that all pages are numbered sequentially—just in case you drop them! If you are using PowerPoint for your talk, consider using the 'View Notes Page' option. This useful option allows you to prepare and print out a set of speaker notes associated with each slide in your show.

If you use a script for a talk, make sure it is written for speaking, not reading.

- *Revise your script.* Put your talk away overnight or for a few days after you think you have finished writing it. Come back to the script later, asking yourself how the talk might be sharpened.

Preparing handouts

- *Consider preparing a written summary for the audience.* In general, an oral presentation should be used to present the essence of some body of material. You might liken the talk to a trailer for a forthcoming movie that presents highlights and captures the imagination. If members of the audience want to know more, they should come along to the full screening of the film (that is, read the full paper). Depending on the circumstances, it may be helpful, therefore, to prepare a full copy or summary of the paper on which the presentation is based to be distributed to the audience. Alternatively, consider preparing a handout that includes key diagrams and points, together with your relevant contact information.
- PowerPoint can be a very helpful tool for preparing a summary of a talk. It includes among its print options a function that allows you to prepare handout copies of images you plan to use during your talk. With the aid of such a document, the audience is better able to keep track of the presentation and you are freer to highlight the central ideas and findings instead of spending valuable time on detailed explanations.

Visual aids

Slide projections, PowerPoint projections, models, white-board sketches, overhead transparencies, video tapes, maps and charts can help to clarify ideas that the audience may have difficulty understanding. They can also help to hold the audience's attention and promote interaction with the audience. Table 8.1 sets out a variety of visual aids and some of their advantages and disadvantages.

Table 8.1 Advantages and disadvantages of various visual aids

Type	Advantages	Disadvantages
Whiteboard or blackboard	• Reinforces main points • Allows for some use of colour • Good for building up a series of connected ideas	• Not good for large and/or complicated diagrams • Not useful for large audiences

Type	Advantages	Disadvantages
	• Easy to organise and can be used outdoors	• You usually have to turn your back to the audience • Requires clear handwriting • Writing during a talk can be time-consuming
Flipchart or paper pad stand	• Inexpensive and easily transported • Important material can be prepared in advance • Prompts can be pencilled in beforehand	• Suitable only for small groups • You usually have to turn your back to the audience • Needs a stable easel for support • Requires clear handwriting • Writing during a talk can be time-consuming
Prepared poster	• Provides a brief striking message • Can include complex colour and design elements	• Can be large and awkward to carry • Can be costly to design and produce
Overhead transparency projector (OHP)	• Images can be seen by everyone • Good for prepared material using colours or diagrams • Can be prepared from computer-generated slides • No need to turn away from audience • Can use overlays • Can be masked so you can reveal information gradually • Can be stored and reused	• Needs a power source • Needs to be correctly aligned • Projector bulbs can fail without warning • Material can be too small to be read • Transparencies can be awkward to manage • Projectors becoming less and less common

(cont.)

Table 8.1 (*cont.*)

Type	Advantages	Disadvantages
Slides (consider saving slides to a CD or computer as an alternative way of using these images)	• Images are of better quality than with OHP • Better at displaying pictures and photographs than OHP • Can be stored and reused	• Projector needs a power source and a darkened room • Needs careful slide alignment and projector focus (projectors are notoriously unreliable) • Expensive to produce
Computer-generated visual with data projector	• Relatively easy to combine text, graphics, audio and video • Can generate a sophisticated project relatively cheaply	• Needs a power source • Needs a computer and projector • Technical difficulties in usage are common
Real object	• May be readily available and convenient to display • Audience can see how object works and looks, and it can be used	• May not be suitable for large groups • Potential for damage to object • Not appropriate for large objects
Model	• Works well for large objects • Gives audience sense of scale and of relationships between elements of the object	• May not be to scale • Detail may not be seen by audience • Potential for damage to model • Potentially costly to produce

Adapted from: Street, Hay and Sefton (2005, pp. 177–8).

Before discussing some of the specifics of preparing and using visual aids, it is worth commenting on one of those that is used most widely. PowerPoint has become an almost universally applied visual aid software for presentations (Higgs & Higgs 2008). It can be a tremendously useful tool, but also brings with it a number of challenges. Obvious disadvantages include the need for access to a computer and projector. Less obvious, perhaps, are some preparation and presentation problems. For some students (and more seasoned presenters), preparing PowerPoint slides can become an end in itself, with speakers spending many hours preparing beautiful projections and less time thinking about what they will say. It is very easy to while away hours on the web looking for images that might illustrate a talk (subject to copyright regulations and with appropriate acknowledgment, of course) and searching for the perfect slide colour combinations. Another common trap is that the default options on PowerPoint encourage the creation of slides that consist almost solely of bullet points. It is important to take care to avoid such problems. Remember that PowerPoint is merely an aid to your talk. As the speaker, you must have full command of the material. Leaving aside the need you might have for specific illustrations to make some points, try to imagine how your talk would be arranged and delivered without PowerPoint. Consider using that as the foundation for your talk, then add the potential of PowerPoint where you believe it might help your audience understand your messages. That said, consider the following when you are preparing a PowerPoint show or any visual aids.

- *Prepare a title slide.* This might set out your name, the title of your talk, your contact details and perhaps an outline of the talk to follow.
- *Make visual aids neat, concise and simple.* Simple and clear illustrations are more easily interpreted and recalled than complex versions. Sloppily produced visual aids suggest a lack of care, knowledge and interest. Visual aids ought to be consistent in their style, but should not be boring.
- *Minimise the amount of text on projections.* Try making no more than five or six points on a screen and make each of those in as few words as possible (say, about six words per point).
- *Produce large and boldly drawn visual aids.* Visuals that can be seen from about 20 metres should be adequate in most cases.
- *Ensure the information shown on screen is easy to read.* Use 18- or 24-point text in a clear typeface on PowerPoint slides. If you hand-write your overhead projector transparencies, use printing rather than cursive writing. Also use upper and lower case text because IT IS MUCH EASIER TO READ THAN BLOCK CAPITALS. Finally, ensure that the text colour you use is easy to read against the background. Some combinations (for example, white text on a yellow background) can be very difficult to read.
- *Use a limited range of colours on projections.* As a starting point, try limiting yourself to three colours. Judicious use of colour on complex illustrative material can help clarify the message. Remember that some colours may evoke certain feelings that add to or detract from the case you are arguing. For example, black can be symbolic of death, green of envy and white of purity (see Box 5.4 for more on this concept).
- *Use illustrations rather than text where possible.* Graphic depictions of information such as line graphs, histograms and pie charts are usually

Visual aids must be large, clear and in highly visible colours.

more effective and more easily understood than tables. However, tables can be useful if they are easy to read. Well-chosen and tasteful cartoons can very effectively communicate a message and help to lighten the atmosphere.

- *Avoid taking graphs or tables directly from a written paper.* These often contain more information and detail than can be comprehended readily in a presentation. Redraw graphs and redesign tables to make the small number of points you wish to convey.
- *Plan to leave a good impression.* If you are using PowerPoint, consider preparing an attractive and relevant final image that can be left on the screen when you have finished speaking and are answering audience questions. It is sometimes useful to restate the title of your talk and your name in this final slide.

Rehearsing

With public speaking, practice makes perfect.

Few people, if any, are naturally gifted public speakers. It is a skill that is developed through experience. People new to public speaking often talk very quickly or too slowly, manage their time poorly by labouring minor points, fail to engage with their audience (through eye contact, for example) or do not know how to operate audio-visual equipment correctly. You can overcome such problems through practice and useful feedback from other students or family members. Do not for one moment believe that you will give a better, more engaging presentation by speaking 'off the cuff', without rehearsal and without notes. Almost without exception, such presentations are woeful. There is no escaping the need for adequate preparation and rehearsal.

Rehearse until there is almost no need to consult your prepared notes for guidance—about ten times ought to do it. The intent is not to commit the talk to memory; instead, rehearsing helps to ensure that you have all the points in the right order and that you have a crystal-clear sense of your talk's key message and its trajectory. This is vital to success. Rehearsing also enables you to practise the pacing and timing of your talk, so you don't, for example, spend 90 per cent of your allotted presentation time discussing 20 per cent of your talk's content! It may also allow you to work out whether you sound boring or arrogant. Importantly, it also allows you to manage your spoken material, handouts and audio-visual resources within the time available (Hay, Dunn & Street 2005, p. 166). Also consider the following points while rehearsing:

- *Speak your material out loud!* Not only may you begin to hear problems of logic in your presentation, but you will also have the opportunity to practise pronouncing some of those specialised disciplinary terms, such as 'existentialist' or 'distanciation', that you might have to employ.

- *Time your rehearsals.* Most novice speakers are stunned to find out how long their presentation takes to deliver, especially when compared with the time it takes to read it quietly, or the time they might believe had elapsed while they were talking. Always match the time available for the talk with the amount of material for presentation. Allow for a few extra minutes to compensate for impromptu comments, technical problems, pauses to gather thoughts or the breath-taking realisation that the audience is not following the tale! If you are using visual aids, also ensure that you allow your audience enough time to make sense of your projections, maps or diagrams. If your talk is too long, decide what material can be removed without affecting the main points. It is better to do the pruning beforehand than to be forced to stop your talk, or to make the revisions in midstream.
- *Make full use of the visual aids to be used in the talk.* Visual aids can consume time rapidly as you move from one medium to another (for example, from a video clip to an online display and then back to PowerPoint). Consequently, pay careful attention to the use of time in the delivery of multimedia presentations.
- *If possible, record one or two rehearsals.* It is often useful to make a video recording of a trial presentation. Video cameras do not 'pull punches' in the same way that as audience that is sensitive to your feelings (such as friends and family). If video is unavailable to you, try an audio recording. Find out if you enliven your talk through variations in pitch, tone and pace, or if you use those annoying saboteurs of a good talk: 'umm', 'er', 'you know', 'like' and 'ah'.

Rehearse your talk using all the prompts and props you plan to employ.

Final points of preparation

As a final step, consider the following questions:

- *Are you dressed and groomed for the occasion?* Although the audience's emphasis should be placed on the intellectual merits of your argument, your appearance may affect some people's perceptions of what you have to say. Looking dishevelled suggests that you have no respect for your audience or for yourself. You might not have to go so far as to wear business attire, but you should look as if you have taken some care about your appearance. As a rule of thumb, consider how your audience is likely to be dressed and, if possible, present yourself at a slightly higher standard.
- *Did you bring water?* No matter how well prepared you are, there is a possibility that you will suffer from 'cotton mouth' during your talk. This

unpleasant affliction causes your mouth to dry out and your tongue to swell to such a size that you can scarcely speak. Water consumed during the talk seems to help!

- *How do the visual aids work?* Be familiar with any equipment and software you will be using. Do not be so unprepared that you must exasperate your audience with questions like: 'How do I switch this projector on?' or 'Can anyone work the DVD player?' You should have checked these things before your presentation. It is your talk!

When it is your turn to speak, take control: of technology, timekeeping, volume and vision ...

- *Are you prepared for technical problems?* Ensure that you have a strategy for dealing with the technical glitches that may happen even to the best prepared presenters. For instance, if you plan on using PowerPoint, take hard copies of your slides in case there are problems with your computer or the data projector. If there is a problem with the equipment that you cannot resolve, ask for assistance, stay calm and keep focused on the key ideas of your presentation.
- *Can the audience see you and your visual aids?* Before your talk, sit in a few strategically placed chairs round the room to see whether the audience will be able to see you and your visual aids clearly. Consider where you will stand while talking and take care to avoid obstructing the audience's view.
- *Is there a clock in the room?* If not, make sure you can see your watch or have some other way of checking the time.
- *Is everything else ready?* Are summaries ready for distribution. Are note cards in order? Is your mobile phone switched off?
- *Are you absolutely clear in your mind the central message you wish to convey?* This is critical to a good presentation. Knowing your message will allow you to project the confidence your audience needs if they are to have faith in what you are telling them. It also means that if something does go wrong, or you 'stall' and lose the plot momentarily, you can take a breath, reflect on your central message and resume your talk with the minimum of fuss. Importantly, too, if you do not have the message of your talk firmly established in your own mind, you are unlikely to be able to let anyone else know what that message is.

How to deliver a successful talk

Audiences support speakers who try to do well.

People in your audience *want you to do well.* They want to listen to you giving a good talk and they will be supportive and grateful if you are well prepared, even if you do stumble in your presentation or blush and stammer. The guidelines outlined here are a target at which you can aim. No one expects you to give a flawless presentation, though you would do well not to follow the model described in Box 8.4!

Wear a dark suit and conventional tie; turn down the lights; close the curtains, display a crowded slide and leave it in place; stand still, read your paper without looking up; read steadily with no marked changes in cadence; show no pictures, use grandiloquent words and long sentences.

Booth (1993, p. 42)

Box 8.4 Directive for lulling an audience to sleep

It will make your presentation more convincing and credible if you remember and act on the fact that the audience comprises *individuals*, each of whom is listening to you. You are not talking to some large, amorphous body. Imagine that you are telling your story to one or two people and not to a larger group. If you can allow yourself to perform this difficult task, you will find that voice inflection, facial expressions and other elements important to an effective delivery will fall into place. Also consider the following strategies:

- *Be confident and enthusiastic.* One of the keys to a successful presentation is your enthusiasm. You have a well-researched and well-prepared talk to deliver. Most audiences are friendly—all you have to do is tell a small group of interested people what you have to say. Try to instil confidence in your abilities and your message. Do not start your talk by apologising for the presentation. If it is so bad, why are you giving it?
- *Look interested—or no one else will be!*
- *Speak naturally, using simple language and short sentences.* Try to relax, but be aware that the presentation is not a conversation in a bar. Some degree of formality is expected. Do not use slang or colloquial language unless you have a specific reason for doing so.
- *Speak clearly.* Try not to mumble and hesitate. This may suggest to the audience that you do not know your material thoroughly. You can sometimes make your speech clearer by slowing the rate of delivery.
- *Speak loudly enough.* Be sure that the most distant member of the audience can hear you clearly. If there is a microphone available, use it. Some of

your audience may be hard of hearing and will appreciate a clear, audible presentation.

- *Engage your audience.* Vary your volume, tone of voice and pace of presentation. Involve the audience through use of the word 'you'; for example, 'You may wonder why we used ...' (Lindsay 1984, p. 55).
- *Use appropriate gestures and movement.* Step out from behind the computer monitor or lectern and move around a little, engaging with different parts of your audience.
- *Try to avoid nervous habits.* Be conscious of distracting behaviour, such as jangling money in your pocket, swaying, pacing back and forth, or saying things like 'umm' a lot, which you may have detected in your rehearsals. Find alternative, good speaking habits.
- *Make eye contact with your audience.* Although this may be rather intimidating, eye contact is very important. Avoid looking at any single person for too long. Instead, engage with one person for a few seconds and then move on to another. Not only does this allow you to gauge audience response but you also will find it keeps members of the audience 'on their toes'.
- *Face the entire audience.* Do not talk to walls, windows, the floor, the ceiling, the blackboard or the projector screen. It is the audience—the entire audience—with which you are concerned. If you are using PowerPoint, you should have a print-out of your slides in front of you, so you do not have to turn your back on the audience to consult the screen.
- *Pay attention to audience reaction.* If the audience does not seem to understand what you are saying, rephrase your point or clarify it with an example.
- *Direct your attention to the less attentive members of the audience.* This may not be as reassuring as focusing your presentation on those whose attention you already have, but it will help you to convey your message to as large a part of the audience as possible.
- *Make unusual words and key terms visible to the audience.* You might include these on a PowerPoint slide or in a handout, or write them on a whiteboard.
- *Avoid writing or drawing on whiteboards for more than a few seconds at a time.* Long periods devoted to the production of diagrams may destroy any rapport you have developed with your audience.
- *Stop talking when you first show a diagram, slide or map.* This is to allow the audience time to study the display. Then, take a moment to familiarise your audience with elements of the visual aid (for example, axis labels), remembering that they have never seen the image before. There is no point

Pay attention to your audience: Can they hear? Can they see? Are they bored? Do they understand?

in telling people what the image means before they have had a chance to work out what it is about (Pechenik 2004, p. 255).

- *Point to the audience's screen, not yours.* Some speakers new to PowerPoint will point with their finger to images on the monitor in front of them believing perhaps that the audience can see what they mean. Direct the cursor or a laser pointer to that part of the screen to which you want the audience to attend.

- *Use a cursor or pointer considerately.* Cursors and pointers allow you to draw attention to selected elements of some slide or image. However, they can be very distracting. Avoid shaking a cursor. Simply slide it to the item you wish to highlight and leave it there. If you are using a laser pointer, draw the audience's attention to the matter you wish to emphasise and then switch the pointer off. Do not circle an item for the full time you are discussing it. This is very distracting. And be aware that pointers magnify the barely noticeable shaking hands of almost any speaker into a splendid and captivating tremble (Pechenik 2004, pp. 255–6). Finally, take care not to point a laser pointer at members of your audience.

- *Be sure that projections are high enough for all the audience to see.* As a rule of thumb, make sure the projection is screened higher than the heads of people in the front row of your audience.

- *When you have finished with an illustration, remove it.* The audience's attention will be directed back at you (where it belongs) and will not be distracted. Do not talk about a topic that is different from the one on your visual display.

- *Switch off noisy machines when they are not being used.* If this is impossible, it may be necessary to speak more loudly than usual to compensate for the whirring of the internal electric fan cooling your slide projector or other aid.

- *Keep to your time limit.* Audiences do not like being delayed, but you should take care not to rush at the end. Last-minute haste may leave the audience with a poor impression of your talk. Watch the time as you proceed. If it is apparent that you will run out of time, let the audience know the key headings you planned to cover and then jump straight to the conclusion. This is likely to be the fastest, most effective and most polished way of summarising the balance of your talk. Of course, if you have rehearsed properly, this problem should rarely occur!

Coping with questions from your audience

Treat question time as seriously as the main part of your talk.

The post-presentation discussion that typically follows a talk allows the audience to ask questions and to offer points of criticism for discussion. This is an important part of the overall presentation that can completely change an audience's response to you and your work. Take care to be thorough and courteous in your response to comments and questions. Here are some useful guidelines:

- *Let the audience know whether you will accept questions in the course of the presentation or after the talk is completed.* Be aware that questions addressed during a presentation may disturb the flow of the talk, may upset any rapport developed with the audience, and may anticipate points addressed at some later stage within the presentation. In general, it is wise to ask the audience to keep their questions until the end of the talk. In the case of group presentations, you should have determined before the talk who will answer questions. You may choose a single person to handle all questions on behalf of the group or, preferably, questions can be answered by the person whose part of the talk covered the material being commented on. It is sometimes helpful to let the audience know that the group has a leader to whom all questions should first be directed. That person can then turn the question over to the group member best qualified to answer it.
- *Stay at or near the lectern throughout the question period.* Question time is still a formal part of the presentation. Act accordingly.
- *Be in control of the question and answer period.* However, if there is a chairperson, moderation of question time is their responsibility.
- *Search the whole audience for questions.* Compensate for blind spots caused by building piles, the rostrum and other barriers.
- *Recognise questions in order.* If possible, take care to receive inquiries from everyone before returning to any member of the audience who has a second question.
- *Repeat aloud those questions that are difficult to hear.* This ensures that you heard the question correctly. Repetition is also for the sake of those members of the audience who also may not have heard the question.
- *Clarify the meaning of any questions you do not understand.* If someone asks a question you do not understand, don't hesitate to seek clarification. It will save time and possible embarrassment, and help to ensure you get to the heart of the audience member's query.

- *Address the entire audience, not just the person who asked the question.*
- *Always be succinct and polite in replies.* There are two reasons you should be courteous even to those who appear to be attacking rather than honestly questioning. First, if the intent of the question has been misinterpreted—with an affront being seen where none was intended—embarrassment is avoided. Second, one of the best ways of defusing inappropriate criticism is through politeness. If, however, there is no doubt that someone is being hostile, keep your cool and, if possible, move closer to the critic. This reduction of distance is a powerful way of subduing argumentative members of an audience.
- *Avoid concluding an answer by asking the questioner if their query has been dealt with satisfactorily.* Argumentative questioners may take this opportunity to steal the limelight, thereby limiting the discussion time available to other members of the audience.
- *Deal with particularly complex questions or those requiring an unusually long answer after the presentation.* If possible, provide a brief answer when the question is first raised before offering to speak with the questioner at greater length after the talk.
- *If you do not know the answer to a question, say so.* Do not try to bluff your way through a problem, as any errors and inaccuracies may call into question the content of the rest of the talk.
- *Difficult questions may be answered by making use of the abilities of the audience.* For example, an inquiry might require more knowledge in a particular field than you possess. Rather than admit defeat, it is sometimes possible to seek out the known expertise of a specific member of the audience. This ensures that the question is answered and may endear you to that member of the audience whose advice was sought. It also lets other members of the audience know of additional expertise in the area. However, remember it is your presentation and this technique should be used in moderation. It is not an easy way of avoiding hard questions!
- *Smile.* It is over!

REFERENCES AND FURTHER READING

Alley, M. 2003, *The Craft of Scientific Presentations. Critical Steps to Success and Critical Steps to Avoid*, Springer Science + Business Media, New York.

A very nicely written book with some very sound advice. The volume contains some thoughts on key mistakes speakers can make (for example, preparing slides that no one reads, ignoring Murphy's Law and losing composure).

Anholt, R.R. 2006, *Dazzle 'Em with Style. The Art of Oral Scientific Presentation*, 2nd edn, Elsevier, Burlington.
A detailed and useful review that covers topics including preparation, structure, visual displays and effective delivery, as well as offering an assessment checklist.

Booth, V. 1993, *Communicating in Science: Writing a Scientific Paper and Speaking at Scientific Meetings*, 2nd edn, Cambridge University Press, Cambridge.
Chapter 2 is an entertaining and informative exhortation to speak well at professional gatherings.

Bowman, L. 2001, *High Impact Presentations: The Most Effective Way to Communicate with Virtually Any Audience Anywhere*, Bene Factum, London.

Campbell, J. 1990, *Speak for Yourself: A Practical Guide to Giving Successful Presentations, Speeches and Talks*, BBC Books, London.
At about 150 pages, this is a comprehensive review on preparing and presenting a talk. The book accompanies a BBC video of the same title.

Courtenay, B. 1992, *The Pitch*, Margaret Gee, McMahons Point, NSW.
This book comprises a series of short articles written by Courtenay—acclaimed author of The Power of One, Tandia and April Fool's Day—for the Australian newspaper. A number of the articles discuss the business of making an oral presentation.

Dixon, T. 2004, *How to Get a First: The essential guide to academic success*, Routledge, London.

Eisenberg, A. 1992, *Effective Technical Communication*, 2nd edn, McGraw-Hill, New York.
Chapter 15 is a lengthy discussion on preparing and giving a talk. Some useful notes on aspects of body language and on writing out a talk are included.

Hay, I., Dunn, K. & Street, A. 2005, 'Making the most of your conference journey', *Journal of Geography in Higher Education*, vol. 29, no. 1, pp. 159–71.
Extends this chapter's discussion to look further at effective delivery and ways of taking advantage of a conference experience to forge valuable professional relationships.

Higgs, C. & Higgs, J. 2008, 'Doing PowerPoint presentations', in *Communicating in the Health Sciences*, eds J. Higgs, R. Ajjawi, L. McAllister, F. Trede & S. Loftus, 2nd edn, Oxford University Press, South Melbourne, pp. 211–18.

Knight, P. & Parsons, T. 2003, *How to Do Your Essays, Exams, and Coursework in Geography and Related Disciplines*, Nelson Thornes, Cheltenham.

Lebrun, J-L. 2010, *When the Scientist Presents: An Audio and Video Guide to Science Talks*, World Scientific, Singapore.
A very good ten-chapter book. The author includes a helpful array of content, including material on preparing a good slide presentation, making the most of software and pointers, and how to deal with troublesome questions. However, he also introduces a helpful analogy, likening the role of the presenter to that of a host to the audience.

Lindsay, D. 1984, *A Guide to Scientific Writing*, Longman Cheshire, Melbourne.
Chapter 3 includes a useful example of an aide-mémoire.

Mohan, T., McGregor, H. & Strano, Z. 1992, *Communicating! Theory and Practice*, Harcourt Brace, Sydney.

Montgomery, S.L. 2003, *The Chicago Guide to Communicating Science*, The University of Chicago Press, Chicago.
Chapter 13 provides a very helpful guide to scientists on giving high-quality talks to both professional and lay audiences.

Pechenik, J.A. 2004, *A Short Guide to Writing About Biology*, Pearson Longman, New York.

Stettner, M. 1992, 'How to speak so facts come to life', in *Writing and Speaking in the Technology Professions. A Practical Guide*, ed. D.F. Beer, IEEE Publications, New York, pp. 225–8.
Includes a helpful discussion on why a talk should cover no more than or no fewer than three main points.

Street, A., Hay, I. & Sefton, A. 2005, 'Giving talks in class', in *Communicating in the Health and Social Sciences*, eds J. Higgs, A. Sefton, A. Street, L. McAllister & I. Hay, Oxford University Press, Melbourne, pp. 176–83.

Turner, K., Ireland, L., Krenus, B. & Pointon, L. 2008, *Essential Academic Skills*, Oxford University Press, South Melbourne.
Chapter 9 provides guidance on preparing and delivering oral presentations, and includes several useful activities intended to develop and hone relevant skills. The chapter also provides some discussion of team presentations and leading discussions.

Van Emden, J. & Becker, L. 2010, *Presentation Skills for Students*, 2nd edn, Palgrave MacMillan, Basingstoke.
Discusses giving an effective formal presentation as well as speaking effectively in seminars and tutorials.

Weeks, C. 2010, *Handy Hints for the Novice Conference Presenter: or How to Avoid Throwing Up, Passing Out or Just Having a Nervous Breakdown in Front of a Live Audience*, The Junction, New South Wales.

As the title suggests, this is a lively and accessible volume. Although aimed at people delivering a conference presentations, this book contains a good deal of useful advice for formal classroom talks.

Williams, L. & Germov, J. 2001, *Surviving First Year Uni*, Allen and Unwin, Crows Nest, New South Wales.
Chapter 9 offers brief and sometimes amusing guidance on presenting with style. The chapter includes thirteen tips for top presentations.

Windschuttle, K. & Elliott, E. 1999, *Writing, Researching, Communicating*, 3rd edn, McGraw-Hill, Sydney.

NINE

Passing exams

Examinations are formidable even to the best prepared, for the greatest fool may ask more than the wisest man can answer.

Charles Caleb Colton (1820)

Key topics

- Why have exams?
- Types of exam
- Preparing for an exam
- Techniques for passing different types of exam

Some people flourish in exams, completing their best work under conditions that others might find highly stressful. Good performance may be a consequence of a person's particular response to stress, but it is more likely the result of good exam technique. This chapter discusses exams, their types and strategies for test success. Though the chapter's focus is on two- to three-hour exams typically held at the end of semester, much of the guidance offered here applies to tests of shorter duration.

Why have exams?

Exams serve three main educational purposes. These are to test your:

- level of factual knowledge
- ability to **synthesise** material, and apply skills and knowledge learned throughout a teaching session
- ability to explain and justify your informed opinion on some specific topic.

These reasons provide some indication of the sorts of things an examiner is likely to be looking for when marking a test. Some tests may seek to fulfil only one of these objectives. For example, some short-answer and multiple-choice tests may only examine your ability to recall information. Others, such as essay questions or an oral exam of a thesis, may serve all three purposes.

Types of exams

Exams in geography and the environmental sciences usually fall into one of four categories (or combinations thereof). These types are set out in Table 9.1.

Table 9.1   Types of exams and their characteristics

Closed-book	
Characteristics	Requires that you answer questions on the strength of your wits, level of knowledge and ability to recall information. Consulting any material in the exam room other than that provided by the examiner for the purposes of the test is not permitted.
Subtypes	• multiple choice • short answer • essay answer
Open-book	
Characteristics	You may consult reference materials such as lecture notes, textbooks and journals during the exam. Sometimes the sources you are allowed to consult will be limited by your examiner.
Subtypes	• exam room • take home

Oral exam (viva voce)	
Characteristics	This is used most commonly as a supplement to written exams, or to explore issues emerging from an Honours, Masters or PhD thesis. You may have to give a brief presentation before participating in a critical but congenial discussion with examiners in which you may be asked to clarify and further demonstrate your understanding of your written work.
Online	
Characteristics	This requires that you complete and submit your exam while online. It usually takes place under secure conditions in a computer laboratory. Still relatively uncommon in geography and environmental studies, it is used increasingly in distance-education and 'flexible' delivery. It may embrace each of the exam types set out above—even the oral exam or viva voce through Skype and other forms of videoconferencing.

Of these forms of exam, the closed-book model is perhaps most common. Consequently, the following discussion focuses on that model. Nevertheless, a good deal of the general advice applies to the other exam types. Some specific guidance on other exam forms is provided at the end of the chapter.

An important component of success in exams is good 'exam technique'. Technique can be broken into two parts: preparation and sitting the exam.

Preparing for an exam

Many students are fearful of exams, literally worrying themselves sick about them. But with adequate and timely preparation, a great deal of the anxiety associated with exams can be moderated and managed to yield the best possible results (Sherratt 2009; Van Blerkorm 2009).

Review throughout the term

This is one of the most difficult things to do in preparing for an exam. It is also very important. Try to review course material as the term progresses, beginning on the very first day of classes. You might review by rewriting lecture notes or keeping up to date with notes from assigned readings. Not only will this help you remember material when the time comes to sit the exam, but it will also make it easier to understand lecture material as it is presented throughout the teaching session. That can be a major benefit when it is time to complete the examination. Several chapters in Orr's (2005) book offer detailed guidance for organising in-term revision. It is perhaps useful to remember Dixon's (2004, pp. 152–3) sage counsel that if:

> ... you have worked out what the key questions are in each course you are taking; have looked at old exam papers and paid attention to the course documentation; have attended and paid attention to your lectures and seminars; have made useful and well-organised notes; and have read around the subject and thought carefully about what your line is on the central issues; then you do not have too much to worry about.

Find out about the exam

Many lecturers will give advance notice of the kind of exam you might expect, allowing you to prepare in the right ways. However, if you do not receive that information, ask teaching staff to let you know what you can expect in the exam in terms of the types of question you may be asked, the time allowed, the materials needed and so forth. See Box 9.1 for more details.

Box 9.1 Questions to ask about your exam

- How long will I have to complete the exam? Am I allowed extra time because I come from a non-English-speaking background or for some other reason?
- Do I have to pass the exam to pass the entire course?
- What is the pass mark?
- How much does the exam contribute to the final grade?
- How many questions are in the exam?
- How many questions do I have to answer? Are any questions compulsory?
- What kind of questions are they (for example, essay, multiple choice or short answer)? Are there either/or options available?
- How are the marks divided among the questions?
- Is there any information provided in addition to the exam (for example, mathematical tables)?
- Will I be examined on material not covered in lectures and tutorials?
- Can I use a calculator or dictionary?
- Am I allowed to bring in any notes or books?
- Where is the exam and when is it being held?
- What is the procedure if I am sick on the day of the exam?

Reduce any anxiety by finding out as much as possible about the exam.

Adapted from: Hamilton (2003).

Seek some direction about how you might focus your supplementary reading and other preparation. Listen for clues from your lecturer about the exam's possible content. Sometimes lecturers will offer thinly disguised hints throughout the teaching term.

If past exam papers are available to you, look at them to get a sense of the likely format of the exam you will sit and the main topics it might cover. Be aware, however, that the style and content of exams may change from year to year, so ask your lecturer if there is likely to be any changes in approach.

Friedman and Steinberg (1989, p. 175) suggest that you should try to anticipate questions and areas that might be covered in the exam. Although this can be helpful, it can also be a risky game. Unless you have been told that the exam will concentrate on specific topics, it is usually better to have a good overview of all the material covered in a class. Broad comprehension means that you should be able to tackle competently any question you encounter. If you narrow the scope of your revisions, you are gambling with your grade.

Revising for a limited range of topics covered in a class is a risky strategy.

Be quite sure that you know exactly *when* and *where* the exam is to be held. Being in the right place at the right time for your exam is your responsibility and not that of academic staff.

Find a suitable study space

Arrange a comfortable, quiet, well-lit and ventilated place where you can study undisturbed. If possible, make sure it is a place where you can lay out papers and books without risk of them being moved. For some people, home is a good place to study, though care needs to be taken to avoid distractions such as the garden, internet, television, the refrigerator or other family members. Also try to avoid studying at the dining-room table unless everyone plans to eat from plates balanced on their knees for the weeks you are studying. For other people, the library is a good study venue, but here you may need to find a quiet corner somewhere to minimise the chances of having your concentration disrupted by friends.

Keep to a study schedule

Once you have found out the dates and times of your exams, prepare a study schedule or calendar. This should allocate specific days or parts of days to the revision of each topic. You may have developed a study routine around class times during semester and find the less structured time allowed for exam revision to be disconcerting. Preparing—and sticking to a study schedule—not only allows you to make good use of your time but, as the date of the exam approaches, it will also remind you of the amount of work you have put into revision.

When you make up your study timetable, think about those times when you work best and schedule your study periods accordingly. Avoid doing something like playing squash or working at the supermarket during the times when you concentrate best. If possible, set up the timetable so you can be doing those sorts of thing when you are feeling mentally flat or slow.

An example of an exam study calendar is shown in Table 9.2. In this example, the fortunate student has only two exams. Each exam contributes the same proportion to the final grade in a subject and the student is performing equally well in both subjects. Reflecting this balance, the student has allocated twelve study periods to environmental management and eleven to geography. Some blocks of time for relaxation, exercise and other day-to-day activities are also set aside. In your own timetable you might also find it helpful to schedule a short break every one or two hours. During these times you can make coffee, listen to a piece of music or do something else that will refresh you for the next study period.

Table 9.2 Example of an exam study calendar

Date	A.M. activity	P.M. activity	Evening
20 November	End of classes	Relax, buy groceries etc.	Get notes in order
21 November	Get notes in order	Environmental Management	Environmental Management (play squash)
22 November	Environmental Management	Environmental Management	Environmental Management
23 November	Geography	Geography	Relax (swim)
24 November	Geography	Geography	Geography
25 November	Environmental Management (yoga)	Environmental Management	Environmental Management
26 November	Environmental Management (run)	Environmental Management	Environmental Management
27 November	Environmental Management	**Environmental Management Exam**	Relax
28 November	Geography	Geography	Geography (play squash)

Date	A.M. activity	P.M. activity	Evening
29 November	Geography	Geography (swim)	Geography
30 November	**Geography Exam**	Celebrate!	Continue celebrating!

If the exams outlined in Table 9.2 were weighted differently from one another or if the student were performing better in one topic than another, it would be advisable to devote extra time to specific topics as appropriate. As noted above, be sure to stick to your revision schedule.

Make a personal study timetable and stick to it.

Get started and maintain a positive attitude

Once you have prepared a timetable and found a place to study, do not procrastinate. Get started. There is no doubt that study is hard work, but the longer you avoid preparing for an exam, the more difficult the task becomes. Do not delay your start because of doubts about your ability to complete the exam. Instead, be positive and have faith in your ability to plan, manage and produce your own success.

In most written exams, you will be required to demonstrate an understanding of the subject material rather than simply regurgitate recalled information. Therefore, your revision should focus on comprehension first and facts second. Be sure you understand what the course was about and the relationships between content and overall objectives.

Concentrate on understanding, not memorising

When you have a grasp of course objectives, you will be in a better position to make sense of content. Having a 'conceptual framework' upon which you can hang substantive material will also allow you to respond to exam questions in a critical and informed manner. To develop this understanding, check the course syllabus, lecture notes, essay questions, practical exercises and textbooks. Try to draw a concept map that indicates the key ideas and relationships between them. You may also find it helpful to condense your notes from, say, thirty pages to six pages and then to one page. Then expand them out again without referring to the original notes. This process of condensing and elaborating on material should help you to develop a good understanding of the subject matter.

Usually it is better to study for meaning than for detail.

Vary your revision practices

To add depth and to consolidate your understanding of course material, use various means of studying for a topic (Barass 2002). Set yourself questions, solve problems, organise material, make notes, prepare simple summary diagrams and read notes thoughtfully.

Practise answering exam questions

Have a go at answering past exam questions under exam conditions. Working within the rigorous time constraints of most exams is something for which you should be prepared. Be sure you understand some of the key phrases and instructions that might appear (for example, 'discuss critically', 'evaluate' and 'compare and contrast'). The glossary at the end of this book may help you to clarify some of these terms. You may also find it helpful to take your trial exam answers to your lecturer to confirm that you are on the right track.

One good reason for practising exams under exam conditions has to do with changing technology. You probably do most of your in-term writing on a computer. In an exam you are likely to be asked to write manually for three hours. Can you do that? What happens to your fingers? Computers allow you to move paragraphs around, check spelling and grammar, and make revisions quickly. Paper and pen do not offer those luxuries, so you have to plan and think about your writing much more carefully.

Seek help if you need it

If there is topic matter you do not understand, ask your lecturer. If you are experiencing emotional, medical or other difficulties that will affect your study, speak to the lecturer, your doctor or a university counsellor. You will not be the only person facing such problems. Most universities have policies and procedures for managing these difficulties, but those strategies generally make it easier to deal with potential underperformance before the exam than afterward.

Maintain your regular diet, sleep and exercise patterns

Good exam preparation includes looking after yourself.

Do not make the mistake of popping caffeine tablets and staying up into the early hours of the morning cramming information into your overtired brain! You run the risk of falling asleep during the exam. Radical changes to your lifestyle are likely to increase levels of stress and may adversely affect your exam performance. If you are in the habit of exercising regularly, keep doing that. Most people find that exercise perks them up, makes learning easier and enhances exam performance. Exercise a little common sense, too. Do not let exam revision time coincide with your conversion from couch potato to trainee marathon runner, for example. You might also consider telling your partner, friends and family that you may be a little more difficult to live with while you are studying!

Avoid cramming late into the night immediately before an exam. That is rather like preparing for a 10-kilometre 'fun run' (one of the world's great oxymorons!) by going for a long run the day before. Instead, if your preparation has gone according to schedule you should be ready to sit the exam. Spend the night before catching up on a few key revision issues, making sure you have all you need for the exam, before relaxing and getting some sleep.

On the day of your exam—if the exam is in the morning—it is vital that you consume something to raise your blood-sugar levels. It is better that to eat something like muesli that is slowly digested than something that will give you a momentary buzz (for example, a chocolate bar) before dumping you on a post-sugar low. If you do not think you can eat anything substantial, try several small, healthy snacks (for example, dried fruit and nuts). Do not face an exam on an empty stomach.

Dress appropriately

For a typical closed-book exam in a lecture hall or university gymnasium, the key is to dress comfortably. Be sure that you will be warm enough or cool enough to function at your optimum level. Feel good about how you look. Your performance may match that feeling.

Pack your bag

Make sure you have your student identification card (in most universities you are required to present your ID in order to sit the exam), pencils or pens, ruler, paper, eraser, watch, lucky charms and a calculator (if required). If it is an open-book exam, take the right books and notes with you. Exam booklets and scribbling paper will usually be provided. If the weather is hot, you might also want to pack a drink. This is particularly important for summer exams in some poorly insulated exam venues. Take extra warm clothing if it is possible that the venue will be cold.

Get to the right exam in the right place at the right time

Be sure you know the scheduled time and venue of the exam. Every year, some people turn up in the afternoon for a morning exam or arrive at the wrong place. Arrive in good time—not too early, not too late. When you are planning your departure for the exam, allow for the possibility of traffic delays, late public transport and bad weather. If you do miss the exam for some reason, see your lecturer *immediately*. You will usually find them in their office for the duration of the exam.

Sitting exams: techniques for passing exams

Before an exam, almost everyone feels tense and keyed up. However, if you have studied effectively and know about the type of exam you will be sitting, the anxiety you feel will probably help you perform at a higher level than if you were quite blasé about the whole affair. Breathe deeply and stride into the exam room with a sense of purpose. You know your stuff and you know what the course was about. Here is the opportunity to prove it!

Make sure you have the entire exam

Always make sure you have the entire exam.

Check that you have all pages, questions and answer sheets. Look on the reverse side of every page of the exam to see if there are extra questions hiding there. On rare occasions, printing or instructional errors may occur. If you believe this may have happened, check with the **invigilator**.

Read the instructions carefully before beginning

Check to see how long you have to complete the exam, which questions need to be answered, in what sections you have some choice of questions to answer, and the mark value of each question. It is wise to *repeat* this process after answering the first question (or several in the case of a multiple-choice or short-answer exam) to confirm that you are doing things correctly. Lecturers find it most disheartening to mark an exam paper by a capable student who has not followed instructions. It is even more upsetting to be that student.

Work out a timetable

Calculate the amount of time you should devote to each question. Time allocations can be calculated on the basis of marks per question, as demonstrated in Table 9.3.

Table 9.3 Exam answering timetable: three-hour exam (180 minutes)

Question 1	5 marks	9 minutes
Question 2	10 marks	18 minutes
Question 3	15 marks	27 minutes
Question 4	20 marks	36 minutes
Question 5	50 marks	90 minutes
Total	**100 marks**	**180 minutes**

No time is assigned in this schedule for short rest breaks to stop, stretch and collect your thoughts, although these helpful activities might be taken up within the time available for some sections. This time budget could be modified usefully by allowing about 10 minutes at the end of the exam to proofread answers. This can be a valuable use of time.

To help with time management, write down the time you need to start each question or section of the exam soon after you get into the exam room. Keep this timetable handy and stick to it. It is most important that you not only allocate time carefully but that you *adhere to your timetable*. It would be very poor practice, in the example under discussion, to spend 30 minutes on question 1. The point is obvious but worth stating: a small amount of extra time spent on any question deprives you of time on others. Time discipline may be difficult, but it is undeniably a key to exam success.

Read the questions carefully before beginning

In closed-book essay examinations, you will usually be given some preparatory time (commonly 10 minutes) in which to read the questions and to make notes on scrap paper. Use this time effectively. Carefully choose the questions you will answer and think critically about their meaning. Take note of significant words and phrases, and underline key words. Jot down ideas that spring to mind as you look over the questions. Use the time as a brainstorming session and record your thoughts immediately. Do not rely on your memory. After several hours of answering an exam paper, you may have completely forgotten the brilliant ideas you had for that last question. Your notes will trigger your memory.

Plan your answers

Do not make the mistake of rushing into your answers. Essay questions in an exam need to be approached in much the same way as essay questions in term time (see Chapter 1). Work out a strategy for approaching your answer to each question. Once the exam has begun, use the ideas you jotted down during reading time as the basis of an essay plan for each answer. Make a rough plan of each answer before you begin writing. This might all be done on separate pages in your examination answer booklet. Not only is a plan likely to give a coherent structure to your answers but, if you do run out of time, the marker *may* refer to the plan to gain some impression of the case you were making. However, take care to distinguish essay plans from final answers in your answer booklet (for example, put a pen stroke through the plan).

Plan your exam.

Begin with the answers you know best

There is usually no requirement for you to answer questions in any particular order. It is often helpful to tackle the easiest questions first to build up confidence and momentum. Further, if you have the misfortune to run out of time, you will have shown your best work.

Answer the questions asked

Surprisingly, the most common mistake people make in exams is not answering the question that was asked, sometimes opting to spew forth a prepared answer on a related topic (Barass 2002, p. 175; Friedman & Steinberg 1989, p. 175). Markers want to know what you think and what you have learned about a *specified* topic. As you will appreciate, the right answer to the wrong question will not get you very far!

Examiners prefer focused, concise and careful answers—not pages of waffle.

Examiners are assessing your level of understanding of particular subjects. Do not try to trick them; do not try using the 'shotgun technique', by which you tell all that you know about the topic irrespective of its relevance to the question; and do not try to write lots of pages in the hope that you might fool someone into believing that you know more than you do. Concentrate instead on producing focused, well-structured answers. *That* will impress an examiner.

Attempt all required questions

It is usually easier to get the first 30–50 per cent of the marks for any written question than it is to get the last 30–50 per cent. In consequence, it is foolish to leave any questions unanswered. In the worst case, make an informed guess. If you find that you are running out of time, write an introduction, outline your argument in note form, and write a conclusion. This will provide the marker with some sense of your depth of understanding and may see you rewarded appropriately.

Grab the marker's attention

People marking written exam papers usually have many scripts to assess. They do not want to see the exam question rephrased or repeated verbatim as the introduction to an essay answer. They probably do not want to read long rambling introductions. Instead, they will want you to capture their attention with clear, concise and coherent answers. Spare the padding. Get to the point.

Emphasise important points

You can stress key points by underlining words and by using phrases that give those points emphasis, such as 'The most important matter is ... ' or 'A leading cause of ... '. You might also find it useful to use headings in essay answers to draw attention to your progression through an argument. Your examiner will almost certainly find headings useful. If used judiciously, bullet or numbered lists and correctly labelled diagrams can also be helpful.

Support generalisations

Use examples and other forms of evidence to support the general claims you make (Dixon 2004, p. 161; Friedman & Steinberg 1989, p. 177). Your answers will be more compelling and will signal your understanding of course material more effectively if they are supported by appropriate examples drawn from lectures, reading and your own experience than if they rest solely on bald generalisations.

Write legibly and comprehensibly

Examiners hate writing that is hard to read. It is very difficult to follow someone's argument if frequent pauses have to be made to decipher hieroglyphics masquerading as the conventional symbols of written English. Try to write legibly. If you have problems writing in a form that can be easily read, write on alternate lines or use printing (rather than running writing) to ensure the examiner is able to interpret your work. Keep your English expression clear, too—errors of punctuation and grammar may divert the examiner's attention from the positive qualities of your work. Many problems of expression can be overcome by using short sentences. These also tend to have greater impact than long sentences.

If you are accustomed to word processing, practise your handwriting (legibly) before a long exam.

Leave space for additions

Begin answers to each question on a new page so that you can add material if time allows. This is particularly important if you have been prudent enough to leave yourself some time for proofreading. Often, people find that they recall material about one question while they are answering another. It is useful to have the time and space to add those insights.

Keep calm

If you find that you are beginning to panic or that your memory has gone blank, stop writing, breathe deeply and relax for a minute or two. A few moments spent this way should help to put you back on track. Do not give up in frustration and storm out of the exam room. Why run the risk of working out a way through a problem *after* you have left the exam room but while the exam is still on?

Don't leave an exam early in despair. Stay and think.

Follow the rules

Whatever you do, do not cheat! Exams are carefully supervised. Copying and other forms of academic dishonesty in examinations do not go unnoticed. Your exam mark, course grade or even your degree may be in jeopardy if you are caught cheating during an exam.

Proofread your answers

Allow yourself time to proofread your answers. Check for grammatical errors, spelling mistakes and unnecessary jargon. You may also find time and opportunity to add important matters you missed in the first attempt at the question.

Sensible exam technique is critical to success. If you follow the advice outlined above, you will have taken some major steps towards a distinguished exam performance.

Be considerate to your fellow students

In the interests of the mental balance of other people sitting the exam and perhaps for your own well-being, try to avoid behaving in a distracting way (for example, shaking your leg, tapping a pencil or drumming your fingers on the desk). Depending on how stressed others in the class are, you may not survive the exam!

Multiple-choice exams

Aside from familiarity with the test material, good results in multiple-choice exams depend, in part, on being conversant with some of the peculiarities of this form of exam. As the following points suggest, there is much more to success in a multiple-choice exam than simply selecting (a), (b) or (c).

- Go quickly through the exam answering all those questions you can complete easily. If there is a question you find difficult, move on and return to it later if you have time.
- Make sure you read all the answer options before selecting one.

Make yourself familiar with the quirks of multiple-choice exams.

- Multiple-choice options usually include a number of completely unrealistic alternatives. Delete those options that are clearly incorrect. Also consider deleting humorous responses, as they are likely to have been included as distracters. This will help you to narrow down your choices.
- Avoid extreme answers. For example, if you were asked to state the correct population of New Zealand in 2006 and the options were: (a) 1.1 million, (b) 4.1 million, (c) 7.6 million, (d) 23 million, you would be well advised to avoid option (d) as it is distinctly larger than the other three options.
- Avoid answers that include absolutes (Burdess 1991, p. 57). In the complex world in which we live, 'never', 'always' and 'no one' are rarely true.
- Avoid answers that incorporate unfamiliar terms (Burdess 1991, p. 57). Examiners will sometimes use technical words as sirens to lure you into shallow waters.

- Do not be intimidated or led astray by an emerging pattern of answers. If, for example, every answer in the first ten questions of the test appears to have been 'b', this does not mean that the next answer ought to be a 'b'. Nor, of course, does it mean that the next answer is not 'b'!
- If there appears to be no correct answer among the options provided, choose the option you judge to be closest to correct.
- If in real doubt about the correct answer, select the longest option. It is often more difficult for an examiner to express a correct idea in few words than it is to express an incorrect one.
- Unless you have been advised that penalties are imposed for giving incorrect answers, answer every question. If you do not know the answer, make an informed guess. If the question has four options, you have a 25 per cent chance of getting the answer right.
- Do not 'overanalyse' a question. If you have completed a multiple-choice test and are proofreading your answers, be cautious about revising your initial response. If you are hesitant about changing your original answer to a different response, leave things alone. You probably got it right the first time. Have faith in yourself.

Oral exams

Formal oral exams are quite rare in geography and the environmental sciences, and hence most people do not get the opportunity to practise them as they might a written or multiple-choice test. Honours and postgraduate students sometimes have oral exams as part of their final assessment. Because of their unfamiliar nature, the oral exam (or **viva voce**) can be quite intimidating. But if you think about it, the oral examination is simply a formalised extension of the sort of discussion you might have had with colleagues, friends and supervisors about the subject you are studying. As such, it should not be too daunting a prospect. Remember, the examiners are real people, too. In some instances, they may be quite nervous about the entire process themselves and particularly about their ability to put you at ease so that a genuine, thoughtful discussion can take place.

To prepare yourself for an oral exam, think about its likely aims. If you do not know the aims of the exam, ask your lecturer or supervisor. Generally, the examiners or discussants will want you to fill in detail that you might not have had the opportunity to include in a written paper or in your thesis. The discussants might also wish to use the viva voce as a teaching and learning forum. They may want to encourage you to think about alternative ways in which you might have approached your topic and to discuss those alternatives with them. If you can, try to look forward to a viva voce as a potentially rewarding opportunity to explore a subject about which you may be the best informed.

As specific advice for completing an oral examination, some of the following points may be useful. (More detailed advice can be found in Murray 2003.)

- If the viva voce is about a thesis or some other work you have written, read through that work taking a few notes on its key points and otherwise refreshing your memory about it. Think about critical matters from each section that you would like to discuss. You will then be better positioned to make the exam proceed as you would like, should the opportunity arise.
- As the date of your viva voce approaches, conduct a search of recent journals to see if any relevant papers have been published. Your examiners may not yet have read those papers and this will provide you with an opportunity to impress them! (Beynon 1993, p. 94)
- Enlist the help of a friend or family member who is well versed in the topic of the exam. Try to think of the sorts of things the examiner might ask you to demonstrate, and practise some brief, clear answers. Box 9.2 sets out some questions commonly asked in oral exams. In trying to make your answers to these questions clear to your learned friend or relative, you may discover areas that need further revision.

Box 9.2 Common questions in oral exams

A number of questions appear regularly in oral exams. Give them some careful thought before the exam.

- Why did you select your research question?
- How does your work connect with existing studies? Does it say anything different? Does it confirm other work?
- How does your work fit into geography or environmental sciences? What makes it a distinguishable part of the discipline's literature?
- Why did you select your methodology? Are there any weaknesses in the approach you adopted?
- What are some of the sources of error in your data?
- What practical problems did you encounter? How did you overcome them?
- Are any of the findings unexpected?
- What avenues for future research does your work suggest?
- What are the particular strengths or weaknesses of your research?
- What have you learned from your research experience? What would you do differently if you had your time again?

- At the exam itself, present yourself in a way that is both comfortable and suitable to the formality of the occasion. As a rule of thumb, think about the type of clothing your examiners are likely to be wearing and dress slightly more formally than they will. Your examiners will also appreciate it if you take your sunglasses off and switch off your mobile phone.
- If you are required to make an introductory statement or to deliver a presentation, find out how much of the total exam time should be dedicated to it. Typically, it might be 10–20 per cent of the total exam's scheduled duration. Focus any presentation you make on the key points of your work, perhaps giving greatest attention to any particularly unusual procedures, results or conclusions.
- Try to display relaxed confidence. Maintain eye contact with your examiners, sit comfortably in an alert position and do not fidget. Remember to speak clearly. Try to formulate brief but clear answers. To this end, it is useful to have considered some of the questions you might be asked before the exam begins.
- If you do not understand a question, say so. Ask to have the question rephrased. It may not be you who has the problem; it is quite possible that the question as phrased does not make sense.

Before a viva voce, prepare your answers to questions you are most likely to be asked.

- Take time to think about your answers. Do not feel obliged to rush a response.
- To give yourself a few extra moments to think about a question, you may find it helpful to repeat the question aloud—but not every time!
- In most exams, as in a discussion, you should feel free to challenge the examiners' arguments and logic, but be prepared to give consideration to their views.
- If it appears to you that the examiners have wrongly interpreted something you have said or written, let them know precisely what you meant.

Open-book exams

The open-book exam is deceptive. At first, you might think, 'What could be easier than a test into which I can take the answers?' Later—perhaps as late as during the exam itself—you may come to realise that these exams can be a trap for the inexperienced. Open-book exams rarely ask for factual answers that you can simply look up. You actually have to know the subject matter very well. The open-book exam simply allows you access to specific examples, references and other material that might support and enhance your answers to questions. You still have to interpret the question and devise an answer. You must produce the intellectual skeleton upon which your answer is constructed, and then place upon that the flesh of personal knowledge, example and argument drawn from reference material. If you do

Take-home and open-book exams require careful preparation.

not understand the background material from which the questions are drawn, you may not be able to perform as well as you should.

The following may assist you in preparing for an open-book exam:

- Study as you would for a closed-book exam.
- Prepare easily understood and accessible notes for ready reference.
- Become familiar with the texts you are planning to use. If appropriate, mark sections of texts in a way that will allow you to identify them easily (for example, using a highlighter pen or sticky notes). Do not mark books belonging to other people or libraries!

Take-home exams

While a take-home exam may appear to offer you the luxury of time, preparation remains a key to success. Do not make the mistake of squandering your time trying to get organised *after* you have been given the exam paper! Not only will you be 'burning daylight' but the preparation also is likely to be hurried and inadequate. Here are some suggestions to help you prepare for a take-home exam.

- As with an open-book exam, a take-home exam allows you the opportunity to produce better answers than might otherwise be possible because you have access to reference materials. So, it is important to have at your disposal appropriate reference material with which you are familiar. In short, you should have read through, taken notes from, and highlighted a sufficiently large range of reference materials to allow you to complete the exam satisfactorily. Arrange all the reference material in a way that allows you to find specific items quickly. For example, arrange materials alphabetically by author, under subject headings, by date of publication or by some other method you find useful.
- Keep to a timetable like the one shown in Table 9.4. The student in this example has decided that two eight-hour working days (for example, 8 a.m. to noon and 1 p.m. to 5 p.m.) are required to complete the exam. Evenings might be spent proofreading answers. Of course, any time allocations you make do need to take account of the other demands you face such as child-care, paid employment, sporting engagements and so on.

Table 9.4 Example of a two-day, take-home exam time budget (16 hours)

Question 1	20 marks	approx. 3 hours
Question 2	30 marks	approx. 4 hours
Question 3	50 marks	approx. 8 hours
Total	**100 marks**	**16 hours**

- Even if you are given 48 hours to complete an exam, you do not have to work on the questions that long. Your aim should be to produce concise, carefully considered, well-argued answers, supported with examples where appropriate. You are not meant to be writing for the entire time. Instead, think carefully and focus clearly on the questions asked.
- If you will be writing the exam on a computer, be sure that you have access to a reliable printer and that you have sufficient supplies of paper and toner!

Set limits on the time and resources you will use in a take-home exam.

Take-home exams present opportunities and challenges. They provide an opportunity to write very good answers to questions. They also offer the challenge of knowing when to stop. It can be very tempting to spend too much time on the exam, gnawing on it as a dog would a bone.

Online exams

Although they are still fairly uncommon, online exams seem destined to be used more widely in geography and environmental studies. Much of the advice about preparing for and completing exams that has been set out earlier in this chapter applies to this form of exam, but there are a few peculiarities and key points that need to be considered:

- Become familiar with your computing and testing environment. As with any other exam, it is helpful to feel as comfortable as possible with the medium or surroundings in which the exam will be conducted. An unfamiliar computing environment can distract you enough that you perform poorly in an exam. It might be argued, too, that there are more opportunities for things to go wrong in a computer-based exam than in other forms of exam. Although many of these problems are beyond your control, you can prepare yourself to respond to difficult situations. This begins by becoming familiar with the computing environment you will be using.
- If you are not using your own home computer, check to see if the machines you will be using are organised in ways with which you are not familiar. If they are, spend some time getting acquainted with the hardware and software. It may also be possible to find a computer in a location that is quiet enough to allow you to concentrate on the exam. If practice exams are available, try them. This experience will allow you to check and confirm your access to the machine and the exam itself. It will also help you uncover and then resolve any problems that may arise before the exam. If no practice exam is available, ask if one can be made available.
- Deal with special needs. If you have any special needs, related perhaps to visual impairment or seating requirements, make these known to your lecturer or tutor well before the exam. It can be very difficult, if not

impossible, for staff to try to accommodate an individual's specific needs as the exam is starting. At the very least, having these things organised ahead of time will minimise distractions and problems that might prevent you from performing well.

Make yourself comfortable with online exam environments: the computers, the passwords, the malfunctions ...

- Know your user-name and password. Most student computing facilities require that you use an account to gain access to the online exam. These accounts can be specific to each student or they can be generic, allowing each student to use the same account to log on. Be sure that you know your current and correct user-name and password. Do not assume that the account you used last year will work this year. Make sure it does. Make sure, too, that you enter your personal identification and password details correctly when you begin the exam. An error here might identify your answers as those of another student.

- Know how to deal with malfunctions. The best way to prevent a computer or software malfunction is to make sure your computer is restarted before you begin the exam. To do this you may need to arrive at the exam room early and, if you are permitted, power down the machine, wait 30 seconds and then restart the computer. Once the machine has started, it is best to not use the computer for anything until you are allowed to start the exam. If you experience a malfunction with software or hardware, record the time of receipt and content of any error messages you see on the screen and report the problem to the invigilator. If possible, do not dismiss the error message. Computer personnel will need to see the messages to resolve the problem.

- Read instructions carefully. It is important that you read all the instructions provided by staff and presented to you by the software. If you do not understand what something means, ask.

- Disregard your neighbours' progress. In many online exams the order in which questions are presented to students may vary. This is sometimes used as a means to prevent cheating. So do not be discouraged if any or all of your neighbours appear to be moving more quickly through the exam than you. It is possible that the software has presented them with different and perhaps less difficult sections of the exam from those that you are completing.

- Work locally. If possible, when answering a free response exam question (that is, an essay-answer or short-answer question), it is a good idea to compose your answer using a word-processing program and to save it on your hard drive or network drive. You can then 'copy and paste' answers into the exam format or software. By saving your work on your hard drive or the network, you will still have a copy of your work should the exam software fail for some reason. Having said this, for security reasons some testing software and computer testing environments will not allow you access to other programs.

- You must also save your work frequently. Make sure that you save about every five minutes if you are actively working on a document. Save more frequently if what you are writing is complex and difficult to reproduce; less frequently if you are not actively writing.
- Keep records. It is a good idea to record the number of each question you have not answered or questions you are not sure about so that you can easily go back to them. This is particularly important in multiple-choice and true/false exams. Most students complete these exams well within the allocated time and often feel impatient about searching through each exam question looking for questions that should be checked. Most exam software allows you to edit and change answers to questions until you choose to submit them for assessment.
- Submit your answers. It is not unknown for students to answer all exam questions and log off without formally submitting their answers for assessment. Make sure you submit your answers!

If permissible, when you finish your exam, save a copy of your answers and get a printout of your work and some acknowledgment of your submission. This might take the form of an on-screen notification of a receipt number. Finally, if the software provides you with a score, record it and keep it so that you can compare it with the final grade you receive.

Good luck!

REFERENCES AND FURTHER READING

Barass, R. 2002, *Study! A Guide to Effective Study, Revision and Examination Techniques*, 2nd edn, Chapman & Hall, London.
Chapters 11 and 12 of this book comprehensively review steps in preparing for and completing various types of examination. Barass's book is a very helpful reference.

Beynon, R.J. 1993, *Postgraduate Study in the Biological Sciences: A Researcher's Companion*, Portland Press, London.

Burdess, N. 1991, *The Handbook of Student Skills for the Social Sciences and Humanities*, Prentice Hall, New York.
Includes two dozen useful pages on exam preparation, conduct and review.

Burns, T. & Sinfield, S. 2008, *Essential Study Skills. The Complete Guide to Success at University*, 2nd edn, Sage, London.
See chapter 12 for useful advice on exam preparation and success.

Cottrell, S. 2007, *The Exam Skills Handbook*, Palgrave MacMillan, Basingstoke.
A useful guide that includes a variety of exercises to help establish your own best approach to exam study.

Dixon, T. 2004, *How to get a First: The Essential Guide to Academic Success*, Routledge, London.

Friedman, S. & Steinberg, S. 1989, *Writing and Thinking in the Social Sciences*, Prentice-Hall, Englewood Cliffs, New Jersey.
Chapter 13 discusses, in some detail, ten handy hints to help improve performance in written exams.

Hamilton, D. 2003, *Passing Exams: A Guide for Maximum Success and Minimum Stress*, new edn, Thomson Learning, London.

Kneale, P. 1997, 'Maximising play time: Time management for geography students', *Journal of Geography in Higher Education*, vol. 21, no. 2, pp. 293–301.

Murray, R. 2003, *How to Survive Your Viva: Defending a Thesis in an Oral Examination*, Open University Press, Maidenhead.
A detailed discussion of the oral examination, including an extensive outline of the kind of questions that might be asked and dealing with 'non verbals' (e.g. issues of body language).

Orr, F. 2005, *How to Pass Exams*, 2nd edn, Allen & Unwin, North Sydney.
A comprehensive book on exam preparation and performance. Most emphasis is given to effective intellectual, physical and psychological means of preparing for examinations.

Sherratt, P. 2009, *Passing Exams for Dummies*, Wiley-Blackwell, Malden, Massachusetts.
A helpful book written in the straightforward and sometimes humorous way that has made the ... for Dummies series of books so popular.

Tracy, E. 2006, *The Student's Guide to Exam Success*, 2nd edn, Open University Press, Buckingham.

Turner, K., Ireland, L., Krenus, B. & Pointon, L. 2008, *Essential Academic Skills*, Oxford University Press, South Melbourne.
Chapter 10 provides guidance on preparing for and completing exams, including multiple-choice, true/false and fill-in-the-blank varieties. It also offers some pointers on how to respond to less-than-favourable results.

Van Blerkom, D.L. 2009, *College Study Skills: Becoming a Strategic Learner*, 6th edn, Wadsworth Cengage Learning, Boston.
This helpful book includes several chapters on preparing for and passing different forms of exam.

Williams, L. & Germov, J. 2001, *Surviving First Year Uni*, Allen and Unwin, Crows Nest, New South Wales.
Chapter 11 offers brief and sometimes amusing guidance on examination preparation and technique.

TEN

Referencing and language matters

If the English language made any sense, a catastrophe would be an apostrophe with fur.

Doug Larson

A man will turn over half a library to make one book.

Samuel Johnson

Key topics

- What are references and why do we need them?
- The author–date (Harvard) system
- The note system
- Plagiarism and academic dishonesty
- Sexism and racism in language
- Some notes on punctuation

One important convention and courtesy of academic communication is citing or acknowledging the work of those people whose ideas and phrases you have borrowed. This chapter provides an outline of the form and practice of the two styles of referencing used most commonly in geography and environmental disciplines. The chapter also discusses the serious matters of plagiarism and sexist and racist language, before concluding with some comments on common punctuation problems.

What are references and why do we need them?

When you use information that has originally appeared in someone else's work, you must acknowledge clearly where you sourced it from. You must always make the acknowledgment in a consistent and recognisable format. Such acknowledgments are called 'references'. In your academic work you are expected to draw on evidence from, and substantiate claims with *up-to-date*, *relevant* and *reputable* sources. The *number* and *range* of references used for your work are also important.

You must cite all references to:

- acknowledge previous work conducted by other scholars
- allow the reader to verify your data
- provide information so that the reader can consult your sources independently.

You must clearly acknowledge your references when you **quote** (use the original source's exact words), **paraphrase** (express a source's ideas in different words) or **summarise** (outline the main points) information, ideas, text, data, tables, figures or any other material that originally appeared in someone else's work. This will help ensure you avoid the risk of plagiarism. References may be to sources such as books, journals, newspapers, maps, films, photographs, reports, electronic sites or personal communications (for example, letters or conversations). References must provide enough bibliographic information for your reader to be able to find your source easily. 'Bibliographic' refers to the key descriptive elements of a publication such as a book, journal article, video or online resource. These elements include details such as author, date of publication, title, volume number and page numbers. In the case of electronic resources, they might also include both the date and point of access (for example, the **URL** or name of database).

Most geographers and environmental scientists use the author–date referencing system. There are three principal systems of referencing: the **author–date system** (sometimes called the Harvard, in-text or scientific system); the **note system** (sometimes called the endnote or footnote system); and the Vancouver system. Of the three systems, the author–date (and variants of it) is the most widely used in geography and the environmental sciences. Although it is less common, some geography lecturers and journals employ the note system. For this reason, the note system is also discussed in the pages that follow. However, because the Vancouver system is confined largely to medicine and other health-related disciplines, it is not discussed here. If you need to employ this approach to acknowledging sources, consult the online guide

Citing medicine: the NLM style guide for authors, editors and publishers (Patrias 2009) for advice. Use whichever system of referencing is recommended by your lecturer and no matter which system you adopt, be sure to employ the same style throughout any single piece of work.

Use one system of referencing consistently throughout a piece of work.

The advice that follows is based largely on the sixth edition of the *Style Manual for Authors, Editors and Printers* (Snooks & Co. 2002). You are advised to consult this reference if you have detailed queries that are not addressed here.

Despite the frequency with which electronic sources (for example, web and email) are now used, there are few comprehensive, consistent and definitive guidelines for citing electronic sources. The sections below suggest formats for a wide range of contemporary sources (for example, satellite images, electronic journals, websites, email and software). Should you have a type of source not included below, adhere to the same fundamental principles that are used in print media referencing. That is, make sure you provide sufficient details for a reader to be able to track down for themselves the source you are citing. Present that information in a sequence that corresponds with the information you provide for your other sources.

The author–date (Harvard) system

The Harvard system comprises two essential components: brief *in-text* references noted throughout your work and a comprehensive *list of references* cited at the end of the work. The in-text reference gives the family name of the *author(s)*, the *date* of publication and the *page(s)* where the information or quotation can be found. The list of references gives the author's family name and forename initials, year of publication, the title of the publication, the name of the publishing house and place of publication, plus additional elements as required (such as chapter name, edition number, and journal name, volume and number). These are discussed more fully below.

In-text references

The in-text reference presents a summary of bibliographic details. Take particular note of the ways in which the various references are punctuated.

- If the reference is to a *single page*:

   As Bloggs (2012, p. 50) has made so clear, a significant challenge confronting geography …

- If the reference is to *several following pages*:

Environmental management is an area of growing employment opportunity (Jones 2011, pp. 506–7).

> To avoid disruption in the flow of the sentence, the citation of author, date and pages used is generally placed at the end of the sentence, although (as in the first example) there are occasions when it is better placed within the sentence.

- If the reference is to a *number of authors*:

Several authors (Brown 2009, p. 9; Henare 2012, p. 16; Nguyen 2010, p. 52) agree …

> Note that each reference is separated by a semicolon. Such lists can be set out in chronological order, reverse chronological order or alphabetically by author name. Whatever approach you adopt, use it consistently through your work.

In-text references provide summary bibliographic details only. Full details of each reference should be included in a separate list of references at the end of your work.

- If the reference is to a single text written by *two or three people*:

A recent study (Chan & D'Ettorre 2009) has shown …
It has been made clear (Brown, White & Green 2008) that …
Chan and D'Ettorre (2009) have shown that …

> Note that the names are linked within parentheses by an ampersand (&) but an 'and' is used if the names are incorporated within the text.

- If the reference is to a single text written by *more than three people*:

Ninio et al. (2011, p. 16) argue that …

> The abbreviation '**et al.**' is short for the Latin term *et alii* meaning 'and others'. Modern convention means that this term no longer needs to be written in italics.

- If the reference is to an *anonymously written work*:

This is apparently not the case in Thailand (*Far Eastern Economic Review* 28 January 2012, p. 12).
Poughkeepsie Yearning (1963) offers fine testament to this view.

> The expressions 'Anonymous' and 'Anon.' should not be used. Instead, the work's title is given.

- If the reference is to a *map*:

Low levels of precipitation are evident through much of central Australia (Division of National Mapping 2007).

- If the reference is to work written by a *committee* or an *organisation*:

CSIRO (2008, p. 41) suggests that soil degradation is of major concern to the agricultural community in Australia.
Natural disasters may present significant difficulties for residents of New Zealand (Earthquake Commission 1995, p. 12).

Occasionally, a publication will have both individual and organisational authors listed. In such cases, it is common practice to treat the individual as author. The organisation is mentioned when giving full details in the list of references cited.

- If the reference is to *one author referred to in the writings of another*:

Motor vehicles are a major cause of noise pollution in urban areas (Hassan, in Yeo 2011, p. 219).

Avoid such references unless tracing the original source is impossible. You are expected to find the original source yourself to ensure that the information has not been misinterpreted or misquoted by the intermediate author.

- If reference is made to information gained by means of *personal communication*:

Aspects of the theory await investigation (Wotjusiak 2010, pers. comm., 2 May).

References to personal communications should be incorporated more fluidly into the text than in the example above and, where appropriate, should include some indication of the quoted person's claim to authority or expertise in the relevant field. For example:

In an interview conducted on 12 January 2012, Dr Melody Jones, Director of City Services, revealed that ...

Personal communications are not usually included in the list of references, but if you have cited a number of personal communications, it may be useful to provide a separate list that provides the reader with some indication of the credibility of the people cited (that is, some indication of their authority in the context of your work).

- If the reference is to *electronic information*:
    The style of reference provided is the same as that for individual, group, organisational and committee authors as outlined above. The in-text reference should indicate the author's name (which may be an institution) and provide details of the year in which the reference was created or last

amended (for websites this is usually noted towards the end of a page as 'Page last updated on ...').

The National Aids Information Clearinghouse (2002) guidelines give clear advice on ...
The full text of David Harvey's (1989) book, *The Condition of Postmodernity*, is now available on the web.

If there is no indication of a production date, you should record this as 'n.d.' for no date. For example:

Genetic engineering seems likely to have a profound influence on Australian crop production figures (Witherspoon n.d.).

The in-text reference for a web resource should *not* include the site or page's URL. This information is only included in the reference list.

List of references

This *alphabetically* ordered (by family name of author) list provides the complete bibliographic details of all sources actually referred to in the text. By convention it does not include those sources you consulted but have not cited (a full list of *all* references consulted is known as a *bibliography*).

The following examples of correctly formatted references may be useful when you prepare your own reference lists. The examples cover a range of commonly encountered sources. Look carefully at the examples and distinguish between the various kinds of work and how each is organised and punctuated. Also note the following:

- The second and subsequent lines of each reference are indented (known as a 'hanging indent').
- Reference lists are single-spaced but with a blank line between each entry.
- Article titles are given minimal capitalisation, but book and journal titles are capitalised.
- Book and journal titles are italicised or underlined. Article and chapter titles are not.
- Article and chapter titles appear in inverted commas.

Article in a journal

Journals are those periodicals typically intended for academic use (Snooks & Co. 2002, p. 204). Examples include *Area*, *New Zealand Geographer* and *Australian Journal of Emergency Management*. Journals commonly give greater emphasis to volume and issue number details than to their day or week of publication (which is stressed with popular magazines).

Stokes, E. 1999, 'Tauponui a Tia: an interpretation of Maori landscape and land tenure', *Asia Pacific Viewpoint*, vol. 40, no. 2, pp. 137–58.

Article in a magazine

Magazines follow a similar style as journals, but the volume and issue number are replaced by details of the publication date.

Brown, A. 2011, 'Muddied waters', *Gourmet Traveller Wine*, 16 January, pp. 23–6.

If the article runs on to additional pages later in the magazine, provide all page numbers. For instance:

Loi, V. 2010, 'Tales from Perth', *Australian Traveller*, 16 January, pp. 23–8, 68.

Complete book, first edition

Oosterveer, P. & Sonnenfeld, D.A. 2011, *Food, Globalization and Sustainability*, Routledge, Abingdon.

Note that all authors' family names precede their initials in a list of references. When multiple authors are listed, no comma is placed between the ampersand (&) and the preceding initial. All authors should be identified in the reference list, even if you have used 'et al.' for works with more than three authors in your in-text reference.

Bibliographic management software (for example, EndNote and ProCite) now makes organising and setting out references easier than ever before.

Complete book, second or later edition

Arbogast, A.F. 2011, *Discovering Physical Geography*, 2nd edn, Wiley, San Francisco.

This is set out identically to the previous reference, except for the addition of the edition number between the book title and the publisher name.

Edited volume

Cope, M. & Elwood, S. (eds) 2009, *Qualitative GIS. A Mixed Methods Approach*, Sage, London.

Most references to edited collections will, in fact, be to specific chapters within the volume rather than to the volume as a whole. For that reason, it is more usual to see references set out in the form of a 'chapter in an edited volume'.

Chapter in an edited volume

Waitt, G. 2010, 'Doing Foucauldian discourse analysis—revealing social identities', in *Qualitative Research Methods in Human Geography*, ed. I. Hay, Oxford University Press, Toronto, pp. 217–40.

Note that the editor's family name and initials are not inverted, and that the title of the book precedes the editor's name.

Government publication

Australian Bureau of Statistics 2011, *Australian Statistical Geography Standard (ASGS): Volume 1—Main Structure and Greater Capital City Statistical Areas, July 2011*, Cat. no. 1270.0.55.001, ABS, Canberra.

Note that there is no full stop after the name of the 'author'. Also, because Australian Bureau of Statistics is given as the author, it can be abbreviated when it appears as the publisher.

Book in a series

Herod, A. 2010, *Scale*, Key Ideas in Geography series, Routledge, Abingdon.

Note that series name appears without italics or inverted commas.

Paper in proceedings

Douglas, F. 1999, 'Sun moths in back paddock', *Balancing Conservation and Production in Grassy Landscapes, Proceedings of the Bushcare Grassy Landscapes Conference, Clare, South Australia*, 19–21 August 1999, Environment Australia, Canberra, pp. 12–21.

Fernando, D. & Bodger, P. 2010, 'Embodied energy analysis of New Zealand power generation systems', *Proceedings of IASTED Technology Conferences*, Banff, Canada, 15–17 July 2010, ACTA Press, Calgary, pp. 235–9.

Be sure your references are punctuated and set out correctly. Many markers (and editors) are very picky about this.

Note that spans of page numbers can be abbreviated, as illustrated in the second example above, but not spans of dates.

Paper in working paper series

Kitchin, R.M. 1997, 'A geography of, for, with or by disabled people: reconceptualising the position of geographer as expert', *SARU Working Paper 1*, School of Geosciences, Queen's University of Belfast, Belfast.

Note that series name operates in the same way as a book title.

Thesis

Ellery, R. 2005, *Landsystems reconstruction of the Wanaka Basin*, MSc thesis, University of Otago, Dunedin.

Marshall, D. 1996, *Putting pokies in place. A consideration of the costs and benefits of pokies since their introduction to South Australia: the case of Peterborough*, BA(Hons) thesis, Flinders University, Adelaide.

Unpublished paper

Fergie, D. 1992, State of racism and racism of the state, unpublished paper presented to the Annual Conference of Museum Anthropologists, in possession of Adelaide University Library, Adelaide.

Morrison, P.S. 2008. On the falling rate of home ownership in New Zealand, report prepared for The Centre for Housing Research, Aotearoa (CHRANZ), New Zealand.

Unpublished materials come in a variety of forms (for example, reports, letters and papers presented at meetings). Ensure that you provide, in a concise way consistent with the style of other references, sufficient details for your reader to be able to gain access to the material cited.

Newspaper

Booker, J. 2011, 'Branson gets quake red-zone tour', *New Zealand Herald*, 20 October, p. 1.

Taylor, I. 2011, 'Climate adviser blasts business lobby over carbon tax', *Sydney Morning Herald*, 1 June, p. 1.

If the article has no obvious author, provide all the details in the in-text citation; there is then no requirement to put the entry in the full reference list. (Note: this applies only to newspaper articles!)

(the *Age* 28 January 2012, p. 2)

... in the *Waikato Times* (14 February 2012, p. 12)

Media release

Dunn, K. 2000, *The geography of racism*, media release, University of New South Wales, 16 October.

Drifter, A. (Minister for Marine Affairs) 2012, *Coastwatch initiative not at sea*, media release, Parliament House, Wellington, 19 March.

Gillard, J. (Prime Minister of Australia) 2011, *Rebuilding after the floods*, media release, Canberra, 27 January, viewed 20 October 2011, <www.pm.gov.au/press-office/rebuilding-after-floods>.

DVD, video, television broadcast and film

Physical Geography—Forces That Shape Our Earth (DVD) 2010, Quantum Leap, Huntingdon.

Vietnam: Impact of Aid on a Developing Nation (video recording) 2011, Classroom Video, French's Forest, NSW.

Shaking the Tree (motion picture) 2006, Big Picture Releases, Sydney.

The type of audiovisual element is placed in parentheses after the title. For television programs, it is helpful to include the precise date of broadcast and the name of the broadcasting network. Also note that episode titles appear within inverted commas.

'The Tree Man from Hamilton' (video recording) 2012, *Landline*, broadcast
    1 February 2012, ABC Television, Canberra.

Map

References to maps are usually set out in the following format:

Mapping organisation and Date *Map title*, Edition number [if appropriate], Scale, Series.

An incorrectly formatted reference list reflects poorly on your work.

For example:

Department of Lands New South Wales 1996, *Wollongong*, 5th edn,
    1:50 000, Topographic series.

Aerial photograph

Aerial photographs can follow a similar referencing pattern to maps, although it is helpful to include a 'medium of the source' statement that makes it clear that the reference is to aerial photography. In some cases, the photograph will not have a formal title describing its location. If this is the case, describe the general location as best you can and present that as the photograph's title. The general format for a reference to an aerial photograph is:

Custodian, State and Year of Photography, *Title* [may be described by photograph's
    general location], Scale (medium of the source), Survey number, Frame number.

For example:

DEHAA (Department of Environment, Heritage and Aboriginal Affairs), South Australia
    1982, *Robe*, 1:40 000 (aerial photography), Survey 2814, Frame 187.

Satellite imagery

The form of reference is usefully set out as:

Custodian and Year of imagery, *Title* [may include or be made up of a reference to
    platform, sensor type and location] (medium of the source), Date, Path, Row.

For example:

ACRES (Australian Centre for Remote Sensing) 2000, *Landsat, 7 ETM+ image of South
    Australia* (satellite imagery), 19 January, Path 97, Row 83.

Airborne imagery

References to airborne images can be set out in the following form:

Custodian and Year of imagery, *Title* [may include sensor type and location] (medium of the source), Date of imagery, Image centre coordinates.

For example:

HyVista Corporation, Australia 2000, *HyMap image of Brookfield, South Australia* (airborne imagery), 3 February, 34° 46′ S, 139° 45′ E.

Electronic journal article (from a full-text database, journal collection or web)

There are no commonly accepted protocols for acknowledging electronic sources. The guidance that follows sets out some ideas that should cover most referencing situations. In all cases, provide enough information to allow the reader to track down the reference you are citing.

A full-text database is one that provides the complete text of a document. References to these and to electronic journal articles are usefully set out in the following format:

Author's family name, Initial(s). Document or article date, 'Title of document or article', *Title of Complete Work*, vol. [if available], no. [if available], page number/s [if available], Date of viewing, Available: <URL> or (Name of database producer/name of database).

Below are some examples:

Arlinghaus, S.L. & Griffith, D.A. 2010, 'Mapping it out! A contemporary view of Burgess's concentric ring model of urban growth', *Solstice: An Electronic Journal of Geography and Mathematics*, vol. 21, no. 2, viewed 24 October 2011, Available: <www.mylovedone.com/image/solstice/win10/ArlinghausandGriffith.html>.

Monmonier, M. 2001, 'Where should map history end?', *Mercator's World*, vol. 6, no. 3, viewed 9 July 2011, (online Gale Group/Expanded).

Delph-Janiurek, T. 1999, 'Sounding gender(ed): Vocal performances in English university teaching spaces', *Gender, Place and Culture*, vol. 6, no. 2, pp. 137–53, viewed 10 July 2011, (online Bell + Howell/Proquest).

Page numbers are included in the Delph-Janiurek example, as the article is available in PDF format (that is, a photographic image of the original document).

Other web documents

References to web documents will typically be set out in the following format:

Author's family name, Initial(s) [or name of organisation responsible for the web page]. Document date, *Title of document*, Name of publishing agency [if appropriate and if it does not repeat author's name], Date of viewing, <URL>.

Here are some specific examples:

Australian Conservation Foundation 2011, *Campaigns*, viewed 24 October 2011, <www.acfonline.org.au/members/member_login.asp>.

Bureau of Meteorology 2011, *Observing Australian Climate Change*, Commonwealth of Australia, viewed 24 October 2011, <www.bom.gov.au/climate/change/observe_climate_change.shtml>.

Smith, J. 2011, *John Smith's Home Page*, viewed 24 October 2011, <www.unibase.com/local/wmtk/template/example2/homepage.htm>.

Computer Support Group Inc. 2011, *International Time Zone Converter*, viewed 25 June 2011, <www.csgnetwork.com/timezncvt.html>.

If the page or site has no author or no date, set the reference out as follows:

The Cook Islands n.d., viewed 25 October 2011, <www.ck>.

Electronic sources present some referencing challenges. Be sure to provide sufficient information about them to allow someone else to find the source.

You should be wary of sites with no acknowledged or clear author. Think critically about the bases you have for believing what you are reading. Is the source credible? Why? For guidance on assessing websites and web pages, see Table 1.1 of this book.

CD-ROM

Material on a **CD-ROM** is dealt with in much the same way as that for comparable hard-copy publications such as books or journals, except that a CD-ROM statement is inserted after the CD's title. For example:

Pawson, E. 1999, 'Remaking places', in *Explorations in Human Geography: Encountering Place* (CD-ROM), eds R. Le Heron, L. Murphy, P. Forer & M. Goldstone, Oxford University Press, Auckland.

Email, discussion lists and newsgroups

These can be set out in the following format:

Author's family name, Initial(s). <author's email address> Message year, 'Subject line from message' (medium of the source), Description of message or statement of list name [if appropriate], <List or recipient address>, Date of message.

Do not include a personal email address in a reference unless you have permission to do so. Here are some examples:

Thomson, K. <k.thomson@massive.com.au> 2010, 'Historical Geographies of South Australia' (email), IAG-list, <iag-list@ssn.flinders.edu.au>, 9 April.

Brown, T. <tim.brown@hellfire.com> 2012, 'Job opening here' (email), Personal email to Melanie Smith, <m.smith@monastic.com.au>, 3 April.

Software programs and video games

Set these out as follows:

Author's family name, Initial(s) [or corporate author] and Date, *Title of program*, Version, Publisher, Place of Publication [if available].

Here are some examples:

Adobe Systems 2011, *Adobe Reader X*, v. 10.1, Adobe Systems, San Jose, California.

Thomson Reuters 2011, *EndNote X5 for Windows*, Thomson Reuters Inc., Carlsbad, California.

Weblogs and wikis

There are no clearly and consistently prescribed formats for acknowledging sources such as weblogs, wikis and podcasts. Just remember to provide sufficient details for a reader to be able to track down for themselves the source you are citing, and present that information in a sequence that corresponds with the information you provide for your other sources. Also think carefully about citing material from these sources, including **Wikipedia**, which may be authored and edited by anyone irrespective of their credentials or expertise. Try setting these out as follows:

Author's family name, Initial(s) [or corporate author] Year of post, 'Title of post', Publisher, Type of post, Date of post, Date of viewing, <URL>.

Here are some examples:

Statistically Speaking 2010, 'Framework for Measuring Wellbeing; Aboriginal and Torres Strait Islander People', Australian Bureau of Statistics, weblog post, 12 March, viewed 28 June 2011, <http://abs4libraries.blogspot.com>.

Earthquakes Wiki 2010, 'Haiti Earthquake', wiki article, 12 January, viewed 28 June 2011, <http://earthquakes.wikia.com/wiki/Haiti_earthquake_2010>.

YouTube and podcasts

For an audio podcast, use 'accessed' instead of 'viewed'. Try setting these out as follows:

Title of video [or podcast] Year, Title of host page [or author name], Type of post, Date of post, Publisher, Date of access [or viewing], <URL>.

Here are some examples:

'Fossil Sharks' 2011, TheScienceShow, podcast, 11 June, Australian Broadcasting Corporation, accessed 28 June 2011, <www.abc.net.au/rn/scienceshow/stories/2011/3240008.htm>.

'Geography Matters' 2011, Joe Beacher, video, viewed 24 October 2011, <www.youtube.com/watch?v=JyhSHDGg-cw>.

Multiple entries by same author

If you have cited two or more works written by the same author, they should be listed in chronological order by date of publication. If they were written in the same year, add lower-case letters to the year of publication in both the reference list *and* the text to distinguish one publication from another (for example, 2012a, 2012b). List the same-year publications alphabetically according to the initial letters of significant words in the reference's title, and assign letters accordingly. The following example illustrates both same-year publications and same-author publications.

Lam, R. 2010a, *A Digest of Water Weeds in South Australia*, Bastian Publishers, Adelaide.

—— 2010b, 'Water weeds in South Australia', *Journal of Water Science*, vol. 66, no. 4, pp. 6–18.

—— 2011, 'Water weed hazards in New Zealand', *Australian and New Zealand Journal of Water Resources*, vol. 13, no. 3, pp. 135–68.

As the example above illustrates, where more than one reference by an author is included, the author's name can be replaced by a two-em rule in subsequent references.

A completed reference list noting some of the sources listed above and prepared according to the Harvard system is demonstrated in Box 10.1.

Box 10.1
Example of a completed reference list (Harvard system)

References

Arbogast, A.F. 2011, *Discovering Physical Geography*, 2nd edn, Wiley, San Francisco.

Arlinghaus, S.L. & Griffith, D.A. 2010, 'Mapping it out! A contemporary view of Burgess's concentric ring model of urban growth', *Solstice: An Electronic Journal of Geography and Mathematics* (online), vol. 21, no. 2, viewed 24 October 2011, <www.mylovedone.com/image/solstice/win10/ArlinghausandGriffith.html>.

Australian Bureau of Statistics 2011, *Australian Statistical Geography Standard (ASGS): Volume 1— Main Structure and Greater Capital City Statistical Areas, July 2011*, Cat. no. 1270.0.55.001, ABS, Canberra.

Booker, J. 2011, 'Branson gets quake red-zone tour', *New Zealand Herald*, 20 October, p. 1.

Bureau of Meteorology 2011, *Observing Australian Climate Change*, Commonwealth of Australia, viewed 24 October 2011, <www.bom.gov.au/climate/change/observe_climate_change.shtml>.

Cope, M. & Elwood, S. (eds) 2009, *Qualitative GIS. A Mixed Methods Approach*, Sage, London.

DEHAA (Department of Environment, Heritage and Aboriginal Affairs), South Australia 1982, *Robe*, 1:40 000, (aerial photography), Survey 2814, Frame 187.

Ellery, R. 2005, *Landsystems reconstruction of the Wanaka Basin*, MSc thesis, University of Otago, Dunedin.

Department of Lands, New South Wales 1996, *Wollongong*, 5th edn, 1:50 000, Topographic series.

Kitchin, R.M. 1997, 'A geography of, for, with or by disabled people: reconceptualising the position of geographer as expert', *SARU Working Paper 1*, School of Geosciences, Queen's University of Belfast, Belfast.

Lam, R. 2010a, *A Digest of Water Weeds in South Australia*, Bastian Publishers, Adelaide.

—— 2010b, 'Water weeds in South Australia', *Journal of Water Science*, vol. 66, no. 4, pp. 6–18.

—— 2011, 'Water weed hazards in New Zealand', *Australian and New Zealand Journal of Water Resources*, vol. 13, no. 3, pp. 135–68.

Morrison, P.S. 2008. On the falling rate of home ownership in New Zealand., Report prepared for The Centre for Housing Research, Aotearoa (CHRANZ), New Zealand.

Oosterveer, P. & Sonnenfeld, D.A. 2011, *Food, Globalization and Sustainability*, Routledge, Abingdon.

Physical Geography—Forces That Shape Our Earth (DVD) 2010, Quantum Leap, Huntingdon.

Statistically Speaking 2010, 'Framework for Measuring Wellbeing; Aboriginal and Torres Strait Islander People', Australian Bureau of Statistics, weblog post, March 12, viewed 28 June 2011, <http://abs4libraries.blogspot.com>.

Stokes, E. 1999, 'Tauponui a Tia: an interpretation of Maori landscape and land tenure', *Asia Pacific Viewpoint*, vol. 40, no. 2, pp. 137–58.

Thomson Reuters 2011, *EndNote X5 for Windows*, Thomson Reuters Inc., Carlsbad, California.

Waitt, G. 2010, 'Doing Foucauldian discourse analysis—revealing social identities', in *Qualitative Research Methods in Human Geography*, ed. I. Hay, Oxford University Press, Toronto, pp. 217–40.

Order your reference list alphabetically by author.

The note system

This system of referencing provides your reader with note or numerical references to a series of footnotes or a list of endnotes. As described below, these notes set out the bibliographic details of each reference cited. For that reason, you (and your lecturer) may consider it unnecessary to include a consolidated list of references at the end of your work. Including a bibliography is a matter of choice in the note system. However, it is worth noting that for the reader of your work, a bibliography can certainly be helpful for locating specific references you have used.

In-text references, footnotes and endnotes

At each point in the text where you have drawn upon someone else's work, or immediately following a direct quotation, place superscript numeral (for example, [3]). This refers the reader to full reference details provided either as a **footnote** (at the bottom of the page) or as an **endnote** (at the end of the document or chapter). If you refer to the same source seven times, there will be seven separate note identifiers within the text that relate to that source. You must also provide seven endnotes or footnotes. This practice is sometimes simplified by providing full bibliographic details in the first endnote or footnote and abbreviated details in the subsequent notes (Snooks & Co. 2002).

The second and subsequent references to a source do not need to be as comprehensive as the first, but they should leave your reader in no doubt as to the precise identity of the reference (for example, if there are two books by the same author, you will need to provide sufficient detail for the reader to work out which text you are noting). In some disciplines, Latin abbreviations such as **ibid**. (from *ibidem*, meaning 'in the same place'), **op. cit.** (from *opere citato*, meaning 'in the work cited') and **loc. cit.** (from *loco citato*, meaning 'in the place cited') are used as part of that abbreviation process, although this practice is far less common today than it was in the past. If your lecturer wishes you to use these Latin abbreviations within your references, he or she will explain them to you. Alternatively, Snooks & Co. (2002, pp. 214–15) provide helpful guidance.

Note that the style of referencing is very similar to the Harvard system. The principal differences are that the author's initial appears before their family name, rather than after it, and that the year of publication appears towards the end of the reference. When using the note system, be aware of the following conventions:

- The first endnoted or footnoted reference to a work must provide your readers with all the bibliographic information they might need to find the work.
- The reference is indented from the note number so that readers can quickly identify the note they are looking for.
- Article titles are given minimal capitalisation, but book and journal titles are capitalised.

- Book and journal titles are italicised or underlined. Article and chapter titles are not.
- Article and chapter titles appear in inverted commas.
- The appropriate page number(s) for the material being cited should be included in each note. The bibliography will record the full page range of any article or chapter drawn from a larger work (for example, an edited collection or journal).

Article in journal

1 E. Stokes, 'Tauponui a Tia: An interpretation of Maori landscape and land tenure', *Asia Pacific Viewpoint*, vol. 40, no. 2, 1999, p. 139.

Article in a magazine

2 A. Brown, 'Muddied waters', *Gourmet Traveller Wine*, 16 January 2005, pp. 23–6.

3 V. Loi 2006, 'Tales from Perth', *Australian Traveller*, 16 January 2005, pp. 23–8, 68.

The second example shows what to do if the article runs on to later pages in the magazine.

Complete book, first edition

4 P. Oosterveer & D.A. Sonnenfeld, *Food, Globalization and Sustainability*, Routledge, Abingdon, p. 102.

Complete book, second or later edition

5 A.F. Abrogast, *Discovering Physical Geography*, 2nd edn, Wiley, San Francisco, 2011, p. 134.

Edited volume

6 M. Cope & S. Elwood (eds) *Qualitative GIS. A Mixed Methods Approach*, Sage, London, 2009.

Most references to edited collections will, in fact, be to specific chapters within the volume rather than to the volume as a whole. For that reason, it is more usual to see references set out in the form of a 'chapter in an edited volume'.

Chapter in an edited volume

7 G. Waitt, 2010, 'Doing Foucauldian discourse analysis—revealing social identities', in *Qualitative Research Methods in Human Geography*, ed. I. Hay, Oxford University Press, Toronto, pp. 217–40.

Government publication

8 Australian Bureau of Statistics, *Australian Statistical Geography Standard (ASGS): Volume 1—Main Structure and Greater Capital Statistical Areas*, July 2011, Cat no. 1270.0.55.001, ABS, Canberra, 2011.

Book in a series

As you are preparing your work, keep full details of all the references you consult. This will make preparing a reference list or bibliography much easier.

9 A. Herod, *Scale*, Key Ideas in Geography series, Routledge, Abingdon, 2010, pp. 23–4.

Paper in proceedings

10 F. Douglas, 'Sun moths in back paddock', *Balancing Conservation and Production in Grassy Landscapes, Proceedings of the Bushcare Grassy Landscapes Conference*, Clare, South Australia, 19–21 August 1999, Environment Australia, Canberra,1999, p. 15.

11 D. Fernando & P. Bodger, 'Embodied energy analysis of New Zealand power generation systems', *Proceedings of IASTED Technology Conferences*, Banff, Canada, 15–17 July 2010, pp. 235–9.

Paper in working paper series

12 R.M. Kitchin, 'A geography of, for, with or by disabled people: reconceptualising the position of geographer as expert', *SARU Working Paper 1*, School of Geosciences, Queen's University of Belfast, Belfast, 1997.

Note that series name operates in the same way as a book title.

Thesis

13 R. Ellery, *Landsystems reconstruction of the Wanaka Basin*, MSc thesis, University of Otago, Dunedin, 2005, p. 46.

14 D. Marshall, *Putting pokies in place. A consideration of the costs and benefits of pokies since their introduction to South Australia: the case of Peterborough*, BA(Hons) thesis, Flinders University, Adelaide, 1996, p. 97.

Unpublished paper

15 P.S. Morrison, On the falling rate of home ownership in New Zealand, Report prepared for the The Centre for Housing Research, Aotearoa (CHRANZ), New Zealand, 2008, p. 35.

16 D. Fergie, State of racism and racism of the state, Unpublished paper presented to the Annual Conference of Museum Anthropologists, in possession of Adelaide University Library, Adelaide, 1992, p. 1.

Newspaper

17 J. Booker, 'Branson gets quake red-zone tour', *New Zealand Herald*, 20 Oct. 2011, p. 1.

18 I. Taylor, 'Climate adviser blasts business lobby over carbon tax', *Sydney Morning Herald*, 1 June 2011, p. 1.

If the article has no obvious author, provide full bibliographic details in the text only:

(the *Age* 28 January 2012, p. 2)

... in the *Waikato Times* (14 February 2012, p. 12)

Media release

Aside from the location of author initials and year of publication, media releases are presented in the same way as set out under the author–date system:

19 K. Dunn, *The geography of racism*, media release, University of New South Wales, 16 October 2000.

20 J. Gillard (Prime Minister of Australia), *Rebuilding after the floods*, media release, Canberra, 27 January 2011, viewed 20 October 2011, ‹www.pm.gov. au/press-office/rebuilding-after-floods›.

21 J. Thwaites (Minister for the Environment, Victoria), *Wild dog management group members appointed*, media release, 30 July 2011.

DVD, video, television broadcast and film

22 *Physical Geography—Forces That Shape Our Earth* (DVD) 2011, Quantum Leap, Huntington.

23 *Shaking the Tree* (motion picture) 2006, Big Picture Releases, Sydney.

Television programs are identified in much the same way as video recordings, except that the precise date of broadcast and the name of the broadcasting network should be included. Also note that episode titles appear within inverted commas.

24 'The Tree Man from Hamilton' (video recording) *Landline*, broadcast 21 September 2011, ABC Television, Canberra.

Map

References to maps are usually set out in the following format:

Mapping organisation, *Map title*, Edition number [if appropriate], Scale, Series, Date.

For example:

25 Department of Lands, New South Wales, *Wollongong*, 3rd edn, 1:50 000, Topographic series, 1996.

Aerial photograph

Aerial photographs can follow a similar referencing pattern to maps, although it can be helpful to include a medium of the source statement that makes it clear that the reference is to aerial photography. In some cases the photograph will not have a formal title describing its location. If this is the case, describe the general location as best you can and present that as the photograph's title:

Custodian, State, *Title* [may be described by photograph's general location], Scale (medium of the source), Survey number, Frame number, Year of photography.

For example:

26   DEHAA (Department of Environment, Heritage and Aboriginal Affairs), South Australia, *Robe*, 1:40 000 (aerial photography), Survey 2814, Frame 187, 1982.

Satellite imagery

The form of reference is usefully set out as:

Custodian, *Title* [may include or be made up of a reference to platform, sensor type and location] (medium of the source), Date, Path, Row.

For example:

27   ACRES (Australian Centre for Remote Sensing), *Landsat, 7 ETM+ image of South Australia* (satellite imagery), 19 January 2000, Path 97, Row 83.

Airborne imagery

References to airborne images can be set out in the following form:

Custodian, *Title* [may include sensor type and location] (medium of the source), Date of imagery, Image centre coordinates.

For example:

28   HyVista Corporation, Australia, *HyMap image of Brookfield, South Australia* (airborne imagery), 3 February 2000, 34° 46' S, 139° 45' E.

Electronic journal article (from a full-text database, journal collection or web)

A full-text database is one that provides the complete text of a document. References to these and to electronic journal articles are usefully set out in the following format:

Author's initial(s) Family name, 'Title of document or article', *Title of complete work*, vol. [if available], no. [if available], page numbers [if appropriate],

Document or article date, Date of viewing, <URL> or (Name of database producer/name of database).

Below are some examples:

29 S.L. Arlinghaus & D.A. Griffith, 'Mapping it out! A contemporary view of Burgess' concentric ring model of urban growth', *Solstice: An Electronic Journal of Geography and Mathematics*, vol. 21, no. 2, 2011, viewed 8 November 2011, <www-personal.umich.edu/~sarhaus/solstice/sum01/flyer.html>.

30 M. Monmonier, 'Where should map history end?', *Mercator's World*, vol. 6, no. 3, 2001, viewed 9 July 2011, (online Gale Group/Expanded ASAP).

31 T. Delph-Janiurek, 'Sounding gender(ed): Vocal performances in English university teaching spaces', *Gender, Place and Culture*, vol. 6, no. 2, 1999, pp. 137–53, viewed 10 July 2010, (online Bell + Howell/Proquest).

Page numbers are included in the example above, as the article is available in PDF format (that is, a photographic image of the original document).

Other web documents

References to web documents will typically be set out in the following format: Author's initial(s) Family name [or name of organisation responsible for the web page], *Title of document*, Name of publishing agency [if appropriate and if it does not repeat author's name], Document date, Date of viewing, <URL>.

Make sure all your references—and particularly web sources—are credible.

Here are some specific examples:

32 Australian Conservation Foundation, *Campaigns*, 2010, viewed 24 October 2011, <www.acfonline.org.au/members/member_login.asp>.

33 Bureau of Meteorology, *Observing Australian Climate Change*, Commonwealth of Australia, 2011, viewed 28 June 2011, <www.bom.gov.au/climate/change/observe_climate_change.shtml>.

34 J. Smith, *John Smith's Home Page*, 2011, viewed 9 November 2011, <www.unibase.com/local/wmtk/template/example2/homepage.htm>.

35 Computer Support Group Inc., *International Time Zone Converter*, 2011, viewed 25 June 2011, <www.cgsnetwork.com/timezncvt.html>.

If the page or site has no author or date, set the reference out as follows:

36 *The Cook Islands*, n.d., viewed 25 October 2011, <www.ck>.

You should be wary of sites with no acknowledged or clear author. Think critically about the bases you have for believing what you are reading. Is the source credible? Why?

CD-ROM

Material on a CD-ROM is dealt with in much the same way as that for comparable hard-copy publications, such as books or journals, except that a CD-ROM statement is inserted after the CD's title. For example:

37 E. Pawson, 'Remaking places', in *Explorations in Human Geography: Encountering Place* (CD-ROM), eds R. Le Heron, L. Murphy, P. Forer & M. Goldstone, Oxford University Press, Auckland, 1999.

Email, discussion lists and newsgroups

These can be set out in the following format:

Author's initial(s) Family name, <Author's email address>, 'Subject line from message' (Medium of the source), List name (if appropriate), <List or recipient address>, Date of message.

Do not include a personal email address in a reference unless you have permission to do so.

Some examples:

38 K. Thomson, <k.thomson@massimo.com.au>, 'Historical Geographies of South Australia' (email), IAG-list, <iag-list@ssn.flinders.edu.au>, 9 April 2010.

39 T. Brown, <tim.brown@hellfire.com>, 'Job opening here' (email), Personal email to Melanie Smith, <m.smith@monastic.com.au>, 3 April 2006.

Software programs and video games

Try setting these out as follows:

Corporate author [or Author's initial(s) and Family name], Date, *Title of program*, Version, Publisher, Place of publication [if available].

Some examples:

40 Adobe Systems, *Adobe Reader X*, v. 10.1, Adobe Systems, San Jose, California, 2011.

41 Thomson Reuters, *EndNote X5 for Windows*, Thomson Reuters Inc., Carlsbad, California, 2011.

Weblogs and wikis

As noted earlier in this chapter, there are no clearly and consistently prescribed formats for acknowledging sources such as weblogs, wikis and podcasts. Just remember to provide sufficient details for a reader to be able to track down for themselves the source you are citing and present that information in a sequence that corresponds with the information you provide for your other sources. Also think carefully about citing material from these sources, including Wikipedia, which may be authored and edited by anyone irrespective of their credentials or expertise.

42 Statistically Speaking, 'Framework for Measuring Wellbeing; Aboriginal and Torres Strait Islander People', Australian Bureau of Statistics, weblog post, 12 March 2010, viewed 28 June 2011, <http://abs4libraries.blogspot.com>.

43 Earthquakes Wiki 2010, 'Haiti Earthquake', wiki article, 12 January 2010, viewed 28 June 2011, <http://earthquakes.wikia.com/wiki/Haiti_earthquake_2010>.

YouTube and podcasts

For an audio podcast, use 'accessed' instead of 'viewed'. Some examples:

44 'Fossil Sharks', *The Science Show*, Australian Broadcasting Corporation, podcast, 11 June 2011, accessed 28 June 2011, <www.abc.net.au/rn/scienceshow/stories/2011/3240008.htm>.

45 'Geography Matters', 2011, Joe Beacher, video, viewed 24 October 2011, <www.youtube.com/watch?v=JyhSHDGg-cw>.

Multiple entries by same author

Second and subsequent references to a source do not need to be as comprehensive as the initial reference. They must simply provide the reader with an unambiguous indication of the source of the material. For example:

1 P. Carter, *Environmental Management for Beginners*, Green Publishing, Darwin, 2011, p. 56.

2 *Age*, 19 Feb. 2012, p. 7.

3 *The Creature from the Black Lagoon* (motion picture) 1997, Scary Movie Releases, Los Angeles.

4 Carter, *Environmental Management for Beginners*, p. 12.

Bibliography

The bibliography at the end of a work using the note system of referencing includes all the works consulted, irrespective of whether they are cited within the text. As distinct from footnotes or endnotes, where the author's initial appears before their family name, the bibliography is traditionally set out in alphabetical order according to authors' family names to make it easier for the reader to find the full details of sources cited in the text. Further it is customary to either remove references to specific pages that may have been cited or, in the case of chapters and journal articles, replace them with the page ranges of the full chapter or article.

A completed bibliography noting some of the sources listed above and prepared according to the note system is demonstrated in Box 10.2.

Bibliography

Abrogast, A.F. *Discovering Physical Geography*, 2nd edn, Wiley, San Francisco, 2011.

ACRES (Australian Centre for Remote Sensing), *Landsat, 7 ETM+ image of South Australia* (satellite imagery), 19 January 2000, Path 97, Row 83.

Australian Bureau of Statistics, *Australian Statistical Geography Standard (ASGS): Volume 1—Main Structure and Greater Capital Statistical Areas, July 2011*, Cat no. 1270.0.55.001, ABS, Canberra, 2011.

Box 10.2
Example of a completed bibliography (note system)

Booker, J. 'Branson gets quake red-zone tour', *New Zealand Herald*, 20 Oct. 2011, p. 1.

Computer Support Group Inc., *International Time Zone Converter*, 2011, viewed 25 June 2011, <www.cgsnetwork.com/timezncvt.html>.

Cope, M. & Elwood, S. (eds) *Qualitative GIS. A Mixed Methods Approach*, Sage, London, 2009.

DEHAA (Department of Environment, Heritage and Aboriginal Affairs), South Australia, *Robe*, 1:40 000, (aerial photography), Survey 2814, Frame 187, 1982.

Department of Lands, New South Wales, *Wollongong*, 3rd edn, 1:50 000, Topographic series, 1996.

Ellery, R. *Landsystems reconstruction of the Wanaka Basin*, MSc thesis, University of Otago, Dunedin, 2005.

Fernando, D. & P. Bodger, P. 'Embodied energy analysis of New Zealand power generation systems', *Proceedings of IASTED Technology Conferences*, Banff, Canada, 15–17 July 2010, pp. 235–9.

Gillard, J. (Prime Minister of Australia), *Rebuilding after the floods*, media release, Canberra, 27 January 2011, viewed 20 October 2011, <www.pm.gov.au/press-office/rebuilding-after-floods>.

Morrison, P.S. On the falling rate of home ownership in New Zealand, Report prepared for the The Centre for Housing Research, Aotearoa (CHRANZ), New Zealand, 2008.

Oosterveer, P. & Sonnenfeld, D.A. *Food, Globalization and Sustainability*, Routledge, Abingdon.

Stokes, E. 'Tauponui a Tia: An interpretation of Maori landscape and land tenure', *Asia Pacific Viewpoint*, vol. 40, no. 2, 1999, pp. 137–58.

Waitt, G. 2010, 'Doing Foucauldian discourse analysis—revealing social identities', in *Qualitative Research Methods in Human Geography*, ed. I. Hay, Oxford University Press, Toronto, p. 217–40.

Notes and note identifiers

Sometimes you may wish to let your reader know more about a matter discussed in your essay or report, but believe that extra information is peripheral to the central message you are trying to convey. An indication to your reader that this supplementary information exists may be provided within the text through the use of note identifiers, such as symbols (for example, *, §, ¶, ‡) or preferably superscript numbers (for example, [5]). If you are using the numerical system of referencing, these notes should be incorporated into the sequence of your references and should not stand apart from it. If you are using the Harvard system, these notes should appear either at the bottom of the relevant pages or at the end of the document (before the list of references) in a separate section headed 'Notes'.

Plagiarism and academic dishonesty

Correct referencing is an important academic and professional courtesy. Failure to fully acknowledge sources of ideas, phrases, text, data, diagrams and other materials is known as **plagiarism** and is regarded widely as a serious transgression of intellectual etiquette. It may lead to the imposition of severe penalties (for example, expulsion from a subject or from university).

Plagiarism, which is derived from the Latin word for 'kidnapper' (Mills 1994, p. 263), is a form of academic dishonesty involving the use of someone else's words or ideas as if they are your own. Plagiarism may occur as a result of deliberate and deceptive misuse of another person's work or as the result of ignorance or inexperience about the correct way to acknowledge other work. Plagiarism can take a number of forms, including:

Plagiarism is regarded very seriously by all universities.

- presenting substantial extracts from books, articles, theses, other published or unpublished works (such as working papers, seminar or conference papers, internal reports, computer software, lecture notes or tapes, numerical calculations and data) or the work of other students without clearly indicating the origin of those extracts by means of quotation marks and references
- using very close paraphrasing of sentences or whole paragraphs without due acknowledgment in the form of references to the original work
- quoting directly from a source and failing to indicate that the material is a direct quote.

Academic dishonesty takes other forms, too, and may include:

- fabricating or falsifying data, or the results of laboratory, field or other work
- accepting assistance from another person in a piece of assessed individual work, except in accordance with approved study and assessment provisions
- giving assistance, including providing work to be copied, to a person undertaking a piece of assessed individual work, except in accordance with approved study and assessment provisions
- submitting the same piece of work for more than one topic unless the lecturers have indicated that this procedure is acceptable for the specific piece of work in question.

Burkill and Abbey (2004) suggest some useful ways to help avoid plagiarism in your work. First, make it a habit to record all bibliographic details on notes from your reading and on photocopied articles and chapters. Second, and similarly, keep full bibliographic details of all material you obtain from the web—not just the URL. Third, correctly cite all sources with your text and provide a full list of references at the end of your work.

Universities regard academic dishonesty as a very serious matter and will usually impose severe penalties. These are usually set out in their official documents and advice to students. New technologies to detect plagiarism are being refined (for example, EVE, MOSS and Turnitin.com), but often a simple Google search will expose clumsy attempts to pass off others' words as one's own. Many universities do not regard ignorance of what constitutes plagiarism as an excuse (see Box 10.3 for some scenarios to provoke your understanding of the issues). So do take the time to become familiar with referencing procedures.

Box 10.3 Thinking about plagiarism

Having read the material in this chapter on plagiarism and taking account of your own university's policies on academic dishonesty, think about the following six scenarios. Discuss them with classmates and your lecturer. No answers are provided. The intention is to heighten your sensitivity to plagiarism and your understanding of your own university's rules on these matters.

- Scenario 1: Michelle has found some really relevant information from the journal, *Geographical Research*. She copies out a couple of sentences word for word and includes quotation marks and the reference. Is this the work of a plagiarist?
- Scenario 2: Phil is searching the internet for inspiration for his essay on 'Flooding in New Zealand's Bay of Plenty'. He stumbles across a website that contains essays written by other students, and finds an essay that is very similar to the title he's been given. He decides to use some of the main ideas in his essay without mentioning where they came from, implying that they are his own ideas. Do you think this is plagiarism?
- Scenario 3: You are working in the Environmental Sciences lab. It's all going well but then some of the equipment breaks. It will now take ages to get the last batch of results. Your friend did the same experiment last week and offers to give you their results. Would just using their results constitute plagiarism?
- Scenario 4: A group of four students has worked together on a fieldwork project and they have each been asked to submit an individual account of the work as an assignment. Three of the students submit work with whole sections that are almost identical. Is this plagiarism?

- Scenario 5: Joe is in a mess. He waits until the very final week that a major assignment is due before he gets down to it. He rather manically grabs some books on planning and constructs a quite substantial amount of his project by piecing together sections from different sources. In-text references to the original sources are given and he uses his own words to link the sections together. Do you think this is plagiarism?
- Scenario 6: A geography student discovers that two of her classmates who are international students are copying a chapter of a book to be submitted as a semester assignment. When she confronts them, she discovers they do not believe that they are doing anything wrong. They claim that in their home country copying (without attribution) for such assignments is acceptable. Does this constitute plagiarism?

Adapted from: Burkill and Abbey (2004, p. 441), Preston (2001, p. 102).

If you have any queries about plagiarism that are not resolved in this book, have a look at Burkill and Abbey's (2004) paper on avoiding plagiarism, consult a good reference style manual, such as Snooks & Co. (2002), or ask your lecturer for advice.

Sexism and racism in language

You may remember from the discussion on essays in Chapter 1 that through writing we can shape the world in which we live. By using **discriminatory language** in our writing and speaking we may unwittingly be contributing to these unacceptable forms of discrimination in society. In consequence, as you prepare an essay, report or talk, try to avoid sexist and racist terminology and ideas. When you have completed your work, read through it—taking account of the advice set out below—to ensure that your language does not unwittingly, unkindly or unfairly discriminate against people. In doing this exercise you may also learn a little about your own attitudes.

Sexist language

Language may be sexist in a number of ways. These include (Snooks & Co. 2002; Eichler 1991, pp. 136–7; Miller & Swift 1989):

- *the use of false generics*—or using words that refer to one sex when both are being discussed, or that encompass women and men when reference is being made to one sex only; for example, talking about 'parents' when only mothers are being considered, or persistently using 'he' as an exemplar.

- *the use of 'man'*—particularly in compounds, verbs and idioms; for example, workman, manhole, manning the ship, and man and the environment.
- *poor use of pronouns*—for example, the use of he, she, him, his or her to refer to any unspecified person or thing that may be male, female or neither (such as a ship, hurricane or country).

Reflect critically on the language you are using to make sure it is neither sexist nor racist.

- *trivialisation*—which usually sees women's activities denigrated and often implies that women behave more irrationally and emotionally than men; for example, using 'office girl' as opposed to 'filing clerk' or stating that women 'bicker' whereas men 'disagree'.
- *stereotyping*—or portraying men or women in ways that emphasise stereotypical characteristics; for example, depicting men as unemotional, uncaring or clumsy; and women as emotional, passive and nimble.
- *generalisation*—or characterising both men and women on the basis of statements that apply to only women or men; for example, an author writing satirically about the US South once said: 'Who are these people? What are they like? Do they have any pastimes besides fighting, hunting, drinking and writing novels? Do they really sleep with their sisters and bay at the moon?' The paragraph might have been repaired as: 'Do the men really sleep with their sisters and bay at the moon? Do the women wear crinolines and stash their whiskey behind the camellias?' (Miller & Swift 1989, p. 48).
- *parallel/nonparallel treatment in nonparallel/parallel situations*—which sometimes takes the form of women having their role defined through their relationship with a man; for example, 'man and wife', 'Mrs Smith, wife of famous Formula 1 driver, George Smith'. Less commonly the reverse applies.

Racist language

Racism is the discriminatory treatment of people on the basis of their race, ethnicity, culture or nationality. It has its basis in a dichotomy between an 'in group' and an 'out group'. Language may be racist in a number of different ways (Snooks & Co. 2002):

- *in-group as norm; out-group as deviation*—where the ethnic status of the in-group is rarely mentioned, whereas that of out-group members is. This often happens in news headlines, for example, 'Greek takes Queensland political position' or 'Japanese gang threat'.

- *in-group as individuals; out-group as group*—where people in the in-group are often described in terms that reflect their individuality (for example, their educational status or age), whereas members of the out-group have their identity outlined only in terms of association with that group.
- *in-group portrayed positively; out-group portrayed negatively*—for example, 'whingeing Pom', 'Kiwi ingenuity' or 'Aussie battler'.
- *in-group uses euphemisms to express actions with regard to out-groups*—for example, 'detainment' of refugees in Australia when, in fact, they appear to have been imprisoned.
- *out-groups described in stereotypical terms*—for example, Vietnamese immigrants to Australia depicted as being nimble-fingered and therefore suited to some forms of clothing manufacture, or Chinese immigrants viewed as having business acumen.
- *ethnic and racial slurs that set the out-group apart from the in-group*—for example, using derogatory names, slurs and adjectives such as 'wog', 'coon', 'convict' or 'nip'.
- *illustrative language representing particular group where illustrations tend to depict the typical person as white, middle-class and of Anglo-Saxon heritage*—for example, 'Mr John Doe' or 'Miss Jane Citizen'.

Racism and sexism are offensive and divisive. Avoid language that contributes to those, and other, forms of discrimination.

Some notes on punctuation

One of the most important skills you should have by the time you have completed your university degree is the ability to communicate clearly. One of the keys to good written communication is correct punctuation (particularly the comma, **full stop**, **ellipsis**, **semicolon**, **colon**, **quotation marks** and **apostrophe**). The following notes are intended to help rectify some of the most common problems of written English. Please take the time to read and absorb them. For more detailed explanations and examples, consult Snooks & Co. (2002).

Comma ,
- breaks up long sentences (e.g. Now there was only standing room in Second Class, the battered yellow coaches were filled to overflowing, and on the curves I could see people on the roofs of the following carriages.)

- shows a pause or natural separation of ideas (e.g. After the recommendations were implemented, further evaluations were conducted.)
- brackets or separates information in a sentence (e.g. The most common, and most easily rectified, problems on essay writing emerge from incorrect acknowledgment of sources.)
- precedes linking words, such as 'but', 'so', 'hence' and 'whereas' (e.g. The aim was to examine sustainability, but the experiment failed.)
- separates information in a list (e.g. The equipment included one inflatable boat, one motor vehicle and a helicopter.)

Full stop .
- ends a complete sentence (e.g. Geography has not always had a smooth ride.)
- ends an abbreviation where the final letter of the abbreviation is not the last letter of the word (e.g. p. for page and ed. for editor)

Ellipsis ...
- indicates that words have been left out of a quotation (e.g. As the report claims, 'There are many factors determining the state of the physical environment ... but the most important is human intervention'.)
- indicates incompleteness (e.g. She went on and on about the rate of inflation, share market movements, currency exchange rates ...)

Semicolon ;
- connects two sentences or main clauses that are closely connected but are not joined with a linking word (e.g. The initial survey revealed a high interest; results showed that further action is appropriate.)
- separates complex or wordy items in a list (e.g. The following factors are critical: the environmental impact statement; the government and union policies; the approval of business and council; and public opinion.)

Colon :
- introduces a list (e.g. The following factors are critical: precipitation, temperature and population.)
- introduces a quotation that comprises a full sentence or number of sentences (e.g. According to Openshaw (1999, p. 81): 'The fundamental technical change that is underpinning the development of the new post-industrial society is the transformation of knowledge which can be exchanged, owned, manipulated and traded.')

Quotation marks (inverted commas) ' ' or " "

- indicate a shorter quotation as part of a sentence (e.g. For our purposes, militarism can be broadly defined as 'a set of attitudes and social practices which regards war and the preparation for war as a normal and desirable social activity' (Mann 1988, p. 166).)
- show the titles of journal articles, as well as chapter names, song titles and poem titles (e.g. Harvey's paper 'Between space and time' is an example of an important contribution to the field.)

Apostrophe '

- indicates contractions in verbs (e.g. I'm, we'll, can't. Note that abbreviations of this sort belong to the informal register and are not usually acceptable in academic writing.)
- indicates possession, as follows:
  - Place the apostrophe at the end of the owner-word, then add a possessive 's' (e.g. The researcher's results [that is, the results of one researcher]; the researchers' results [that is, the results of more than one researcher]).
  - If the original word ends with an 's', place the apostrophe at the end of the owner-word without adding a possessive 's' (e.g. The thesis' results [that is, the results of a thesis]).

Apostrophes are one of the most commonly misused or neglected elements of punctuation.

It is important to distinguish between it's (= it is) and its (= belonging to it, whether singular or plural).

Capital letters

- should be used minimally, especially in titles and headings (note that small words such as 'and', 'in', 'the' and 'by' should not take **capital letters**)
- should be used only for a specific and formally named item (e.g. France or English)

Numerals

- should not be used at the beginning of a sentence; rather the number should be spelled out (e.g. 'Nine hundred workers were laid off', not '900 workers were laid off')
- should be used for numbers greater than nine, with words used for numbers less than 10, except when followed by units of measurement (e.g. 'nine field sites' but '9 millimetres')
- should have a thin space between the numeral and the unit of measurement
- should not include full stops when indicating units of measurement

REFERENCES AND FURTHER READING

Burkill, S. & Abbey, C. 2004, 'Avoiding plagiarism', *Journal of Geography in Higher Education*, vol. 28, no. 3, pp. 439–46.

Eichler, M. 1991, *Nonsexist Research Methods: A Practical Guide*, Routledge, London.
> This book provides a comprehensive review of the ways in which research practices can be sexist. It also includes a discussion of sexism in language (chapter 7). Many universities now include helpful advice on non-sexist and non-racist writing on their websites.

Harrison, N. 1985, *Writing English: A User's Manual*, Croom Helm, Sydney.
> Chapter 6, 'Making the text live', is an extensive review of the uses of punctuation.

Miller, C. & Swift, K. 1989, *The Handbook of Non-Sexist Writing for Writers, Editors and Speakers*, 2nd edn, Women's Press, London.
> Though dated, this is an extensive, fascinating and useful review of sexist language, its influence, and ways to avoid it. Well worth reading.

Mills, C. 1994, 'Acknowledging sources in written assignments', *Journal of Geography in Higher Education*, vol. 18, no. 2, pp. 263–8.

Patrias, K. 2009, *Citing medicine: the NLM style guide for authors, editors, and publishers*, viewed 20 October 2011, <www.nlm.nih.gov/citingmedicine>.

Peters, P. 1985, *Strategies for Student Writers: A Guide to Writing Essays, Tutorial Papers, Exam Papers and Reports*, Wiley, Brisbane.
> Chapter 9 includes a detailed review of punctuation.

Preston, N. 2001, *Understanding Ethics*, 2nd edn, Federation Press, Sydney.

Snooks & Co. 2002, *Style Manual for Authors, Editors and Printers*, 6th edn, John Wiley & Sons Australia, Canberra.
> At the time of writing this was the most recent official Australian government style guide. It is the updated edition of the 1994 Australian Government Publishing Service's Style Guide.

Finally, most universities now provide online resources that set out institution-specific guidance on referencing, academic honesty and academic language. Check your own institution's website for advice.

GLOSSARY

abstract

a library resource that lists articles from periodicals under subjects and includes a summary of the article: a short statement outlining the objectives, methods, results and central conclusions of a research report or paper. This latter form of abstract is limited in its length (usually about 100–250 words) and is designed to be read by people who may not have the time to read the whole report, who wish to get a quick impression of the paper's content, or who are working out whether the content is sufficiently interesting for them to read the entire document. (See also *informative abstract* and *indicative abstract*.)

academic journals

see *scholarly journals*.

account for

to explain how something came about and why.

acknowledgment

a statement recognising the people and institutions to which an author is indebted for guidance and assistance. It may be incorporated into the preface or foreword. (See also *citation*.)

acronym

a word made up of the first letters of a group of words (for example, SCUBA—Self-Contained Underwater Breathing Apparatus).

analyse

to explore the component parts of some phenomenon in order to understand how the whole thing works. It can also mean to examine closely. (Contrast with *synthesise*.)

anecdote

a story often personalised to the author or presenter, and directly related to the point of the paper or presentation, that captures the attention of an audience and persuades them of the importance, relevance and/or interest of what they are reading or hearing.

annotated bibliography

a list, in alphabetical order by each author's surname, of works (books, papers etc.) on a specific topic. Each work is summarised and commented upon.

annotation

an explanatory note or comment. It may be used as a label for clarification on a graph or other figure.

apostrophe

a punctuation mark that is used to indicate possession (for example, the girl's book or the three boys' books) or the omission of a letter or letters from a word (for example, can't, you'll or we'd).

appendix

supplementary material accompanying the main body of a paper, book or report. Typically placed at the back of the document, it includes supporting evidence that would detract from the main line of argument in the text, or would make the body of the text too large and poorly structured.

appraise

to analyse and judge the worth or significance or something.

argue

see *argument.*

argument

a debate that involves reasoning about all sides of an issue and offering support for one or more cases. Typically, you will be asked to present a case for or against a proposition, presenting reasons and evidence for your position. In an argument you should also indicate opposing points of view and your reasons for rejecting them. An argument may be written or spoken.

arithmetic scale

the scale on a graph where the main values are separated by constant values (for example, 0, 10, 20, 30 ...; or 0, 30, 60, 90 ...). (Contrast with *logarithmic scale.*)

assess

to conduct an evaluation, investigating the pros and cons or validity of some issue or situation. You are usually expected to reach some conclusion on the basis of your research and discussion (for example, is some situation under consideration right or wrong, or fair or unfair?).

author–date system

a system of referring to texts cited that comprises two parts: in-text references, which provide a summary of the bibliographic details of the publication being cited (which provides author's surname, year of publication and page references); and an alphabetically ordered list of references (which provides complete bibliographic details of all sources referred to in the text). (Compare with *note system.*)

bar graph

the general name given to those graphs in which plotted values are shown in the form of one or more horizontal or vertical bars (column graph) whose length is proportional to the value(s) portrayed. (Contrast with *histogram.*)

bibliographic details

information about a publication that includes, but is not limited to, information such as: who is the author; when was the work published; where; and by whom? Depending on the kind of publication, bibliographic details may also include volume, issue and page numbers.

bibliography

a complete list of works referred to or found useful in the preparation of a formal communication (for example, essay, book review, poster or report). Less commonly, a bibliography refers to a book listing works available on a particular subject. (See also *reference list* and *annotated bibliography*.)

blog

a web application that allows single or multiple authors to post news, information, musings, personal diaries, political commentary, photographs etc. on a common web page. Many blogs have little formal quality control and are a medium for expressing personal opinion. Accordingly, they should be used with caution as information sources. Blog is a contraction of the term 'weblog'.

cadastral map

a specialised map showing surveyed land tenure (from *cadastre*, an official register or list of property owners and their holdings).

capital letters

used at the beginning of sentences, and for proper names and titles (for example, On Monday, Ms Jane Smith went home).

caption

explanatory material printed under an illustration. It may also refer to the title or heading above a map, figure or photograph.

cartogram

a form of map in which the size of places depicted is adjusted to represent the statistics being mapped. For example, if one was to produce a map of the world showing sheep populations, New Zealand and Australia would appear to be very large compared with most other countries. Although the physical sizes of places will be altered in the production of a cartogram, efforts are made to preserve both their locations relative to other places and their shapes.

cartographic scale

see *scale*.

CD-ROM

(Compact Disc Read Only Memory) a compact disc that can store large amounts of information, such as text, music and graphics.

choropleth map

a cross-hatched or shaded map used to display statistical distributions (for example, rates, frequencies or ratios) on the basis of areal units such as nations, states and regions.

circle graph

see *pie graph*.

citation

a formal, written acknowledgment that you have borrowed the work of another scholar. Whenever you quote verbatim (that is, recite word for word) the work of another person, and when you borrow the idea(s) of such people, you must acknowledge the source of that information using a recognised referencing system.

class interval

a cluster of data into which the values of a frequency distribution may be grouped.

climograph

a graphical depiction of temperature and precipitation at a specific location.

clincher

that part of a paragraph that concludes the paragraph's argument. (See also *topic sentence* and *supporting sentence*.)

colon

a punctuation mark (:) that means 'as follows' and is used to indicate lists and examples (for example, 'We read: books, articles, essays and magazines').

comment

to make critical observations about the subject matter.

common knowledge

information that is widely known in the community. Matters of common knowledge may be referred to in scholarly work without the need for a source to be cited (for example, Australia is a large island nation west of Aotearoa/New Zealand).

compare

to discuss the similarities/differences between selected phenomena (for example, ideas or places). Be quite sure that you know what you are meant to be comparing. (Often used in conjunction with *contrast*.)

concept

a thought or idea that underpins an area of knowledge. For example, the concept of evolution underpins much of biology, while the idea that new communications technologies 'compress' distance is significant in geography.

conceptual framework

the logic that underpins an argument or the way in which material is presented; also a way of viewing the world and of arranging observations into a comprehensible whole. It may be imagined as an intellectual skeleton upon which flesh in the form of ideas and evidence are suspended.

conclusion
that part of a talk, essay, poster or report in which findings are drawn together and implications are revealed.

consider
to reflect on; think about carefully.

continuous data
observations that could have any conceivable value within an observed range. Thus, they include fractional numbers, such as halves and quarters. (Compare with *discrete data*.)

contrast
to give a detailed account of differences between selected phenomena. (Often used in conjunction with *compare*.)

corroboration
the support or confirmation of an explanation or account through the use of complementary evidence. (Compare with *replication*.)

critical reading
reading a text with the intent of questioning its content and the ideas that lie behind it rather than accepting it at face value.

criticise
to provide some judgment on strengths and weaknesses; to back your case with a discussion of the evidence. Criticising does not necessarily require you to condemn an idea.

critique
see *criticise*.

database
a large amount of information stored in a computer and organised in categories to facilitate retrieval.

data region
that part of a graph within which data is portrayed (usually bounded by the graph's axes).

define
to explain the basic points or principles of something to provide a precise meaning. Providing examples may enhance your definition.

demonstrate
to illustrate and explain by use of examples.

dependent variable
a study item with characteristics that are considered to be influenced by an *independent variable*. For example, flooding is heavily dependent on rainfall.

describe

to outline the characteristics of some phenomenon without necessarily interpreting it. Usually, a description might be imagined to be a picture painted with words. What does the phenomenon look like? What patterns are evident? How big is it?

discrete data

phenomena that may be quantified in whole numbers only; for example, animal and human populations. (Compare with *continuous data*.)

discriminatory language

language that treats people differently, or excludes them, on the basis of gender, race, religion, culture, age disability or sexuality where it is unnecessary or inappropriate to do so. (Contrast *non-discriminatory language*.)

discuss

to examine critically, using argument; to present your point of view and that of others. It may be written or spoken.

distinguish

to make clear any differences between two or more phenomena.

DOI

Digital Object Identifiers, which are unique alpha-numeric strings used to identify documents and other objects (for example, images, books and song lyrics) in the online environment. Whereas a *URL*—or web address—specifies an online location, a DOI specifies content, and is therefore an enduring identifier of online content, regardless of the object's web address.

domain name

a unique name that locates an organisation or other entity on the internet. It is always broken into two or more parts, separated by dots. For example, <iag.org.au> is the domain name for the Institute of Australian Geographers and <nzgs.co.nz> is that for the New Zealand Geographical Society.

dot map

a map in which spatial distributions are depicted by dots representing each unit of occurrence (for example, one dot represents one person) or some multiple of those units (for example, one dot represents 1000 sheep).

edit

to revise and rewrite. It can also imply that some material needs to be deleted.

electronic mail

see *email*.

ellipsis

a punctuation mark, written as three full stops (...). When placed in a quotation, it indicates that words from the original source have been omitted. It can also indicate incompleteness.

email

the exchange of computer-stored messages by telecommunications.

embargo

a request by the author of a media release that their story not be published until after a specified date or time or unless certain conditions have been met. It is often used if information is politically or commercially sensitive.

endnote

a short note placed at the end of a document and identified by a symbol or numeral in the body of the text. A textual 'aside', the endnote provides a brief elaboration of some point made in the text but whose inclusion there might be inappropriate or disruptive to the flow of text. In the *note system* of referencing, endnotes may also include details of reference material cited in the text. (Compare with *footnote*.)

EndNote

proprietary software for managing and producing reference lists and bibliographies. It's a form of *reference management software*.

enumerate

to list or specify and describe clearly.

essay

a brief literary composition that states clearly what you think and have learned about a specific topic.

essay plan (or essay outline)

a series of headings and perhaps some brief explanatory text setting out the basic structure and argument of an essay.

et al.

an abbreviation of the Latin phrase *et alii*, meaning 'and others'. It is used to reduce a list of people's names in citations.

ethics

a framework for the moral conduct of researchers and their responsibilities and obligations to those involved in the research.

evaluate

to appraise the worth of something; to make a judgment. What are the strengths and weaknesses and which are dominant?

evidence

information used to support or refute an argument or statement. In forming an opinion or making an argument at university, you may need to abandon some practices that may have been considered satisfactory in the past. For example, it is not acceptable for you to state such things as 'It is widely known that ...' or 'Most people would say that ...' since in these statements you have not provided any evidence about who the people are, why they say what they do, how they came to their conclusions, and so on. In other words, you need to present material that supports or refutes your claim.

examine

to investigate critically; to present in detail and critically discuss the implications.

explain

to answer 'how' and 'why' questions; to clarify, using concrete examples.

extrapolate

to estimate the value of some phenomenon beyond the extent of known values. Typically, this is done by extending historically known trends into the future. For example, if house prices in Invercargill had been increasing at an average rate of 5 per cent per year for the last 20 years and the median value of a house at the end of last year was $200 000, you might extrapolate from the trend to suggest that the median Invercargill house value will have risen to $210 000 by the end of this year. (Compare with *interpolate*.)

factoid

a 'fact' that is considered to be of dubious origins or accuracy.

first-person narrative

in the context of a research narrative or other communication (for example, a research report, essay or talk), this is the presentation of the researcher/author/speaker as 'I' or 'we'. In first-person narrative, researchers or 'narrators' are able to insert their stance, beliefs, emotions and assumptions explicitly into the text. In so doing, they arguably become more accountable, and their knowledge can be better 'situated'. (Compare with *third-person narrative*.)

footnote

a short note placed at the bottom of a page and identified by a symbol or numeral in the body of the text. A textual 'aside', the footnote provides a brief elaboration of some point made in the text but whose inclusion there might be inappropriate or disruptive to the flow of text. In the *note system* of referencing, footnotes may also include details of reference material cited in the text. (Compare with *endnote*.)

foreword

a message about the main text of a book, it is usually disconnected from that text because it is written by a different author or because it does not contribute directly to the textual content. An example of a foreword might be a statement by a prominent politician about the timeliness and value of the published volume in which the foreword appears. (Compare with *preface*.)

freewriting

sometimes used as a step in the production of an essay. It involves 'stream of consciousness' writing without concern for overall structure and direction, followed by careful revision.

full stop

a punctuation mark used to indicate the end of sentences and some abbreviations.

generalisation

a comprehensive statement about all or most examples of some phenomenon made on the basis of a (limited) number of observations of examples of that phenomenon.

Harvard system

see *author–date system*.

heading

a phrase or word used at the beginning of a section of text (for example, in an essay or report) that encapsulates the content of that section and, with other headings in the document, allows a reader to gain some sense of the structure or intellectual trajectory of the work.

histogram

a graph in which plotted values are shown in the form of horizontal or, more commonly, vertical bars whose area is proportional to the value(s) portrayed. Thus, if class intervals depicted in the histogram are of different sizes, the column areas will reflect this. (Contrast with *bar graph*.)

home page

the main or 'front page' of a website. It usually sets out the content and other characteristics of the site. (See also *website* and *web page*.)

hypothesis

a supposition or trial proposition used as a starting point for investigation. It usually begins with the word 'that'; for example, 'that Mount Ruapehu's most recent eruption promoted tomato growth in horticultural regions of New Zealand's North Island'.

ibid.

an abbreviation of the Latin word *ibidem*, meaning 'in the same place'; sometimes used in footnotes and endnotes.

illustrate

to make clear through the use of examples or by use of figures, diagrams, maps and photographs.

independent variable

a study item with characteristics that are considered to cause change in a dependent variable. For example, the independent variable rainfall may promote flooding (the *dependent variable*).

index

an alphabetical list of names and subjects, with page references, at the back of a book. It is also a library resource that lists citations to periodicals, articles and other information alphabetically, according to subject.

indicate

to focus attention on or point out.

indicative abstract

a short written statement that outlines the contents of a paper, report or book, but does not recount specific details. It is commonly used to summarise particularly long reports and book chapters. (Compare with *informative abstract.*)

informative abstract

a short written statement that summarises a longer account of primary research (for example, a paper or report), offering a concise description of the work's content, including aims, methods, results and conclusions. (Compare with *indicative abstract.*)

internet

a worldwide network of computer networks that enables communication between computers connected to the network. (Also see *World Wide Web.*)

interpolate

to estimate a value of some phenomenon between, and on the basis of, values that are already known. For example, if you knew that the median price of a house in Mount Gambier was $210 000 in January and $220 000 in December, you might interpolate the June value to have been about $215 000. (Compare with *extrapolate.*)

interpret

to make clear, giving your own judgment; to offer an opinion or reason for the character of some phenomenon.

introduction

the first section in a piece of formal communication (for example, poster, talk or essay) in which author or speaker tells the audience what is going to be discussed and why.

inverted comma

see *quotation marks.*

invigilator

a person who supervises the conduct of examinations. This may not be the same person who teaches the course.

isoline map

a map showing sets of lines (isolines) connecting points of known, or estimated, equal values. Common examples include topographic maps, which show lines of equal elevation (contours), and weather maps, which commonly show isobars (lines of equal atmospheric pressure).

jargon

most commonly, technical terms used inappropriately or when clearer terms would suffice. Less commonly, it is words or a mode of language intelligible only to a group of experts in the field.

journal

a publication issued at regular or irregular intervals on an ongoing basis (for example, *Australian Journal of Environmental Management* or *New Zealand Geographer*). Also called periodicals, magazines or serials.

justify

to provide support and evidence for outcomes or conclusions.

key

see *legend*.

legend

a brief interpretive statement making sense of the symbols, patterns and colours used in a map or diagram. Also known as a key.

line graph

a graph in which the values of observed (x,y) phenomena are connected by lines. It is used to illustrate change over time or relationships between variables.

line of best fit

a straight line that best summarises the data on a scattergram. The line may pass through some, all or none of the points on the graph.

literature review

a comprehensive summary and interpretation of resources (for example, publications or reports) and their relationship to a specific area of research.

loc. cit.

an abbreviation of the Latin phrase *loco citato*, meaning 'in the place cited'. It is sometimes used in footnotes and endnotes.

logarithmic graph

a graph with one (*semi-log*) or two (*log–log*) logarithmic axes. (Also see *logarithmic scale*.)

logarithmic scale

a scale on a graph axis in which the key elements are based on exponents of 10 (for a base-10 scale). The base-scale may be different, but 10 is used most commonly. Logarithmic scales are most useful when data span large ranges (for example, gross national product). (Contrast with *arithmetic scale*.)

log–log graph

graph with logarithmic scales on both axes. (Compare with *semi-log graph*.)

map

a graphic device that shows where something is; a graphic representation of a place.

media release

structured information about some issue or event provided in textual or visual form to a media outlet (for example, television station or newspaper).

narrate

to say what happened in the form of a story.

newsgroup

an online discussion group. Notices and messages often follow a particular theme or topic (known as a thread) within the broader interests of newsgroup members.

non-discriminatory language

forms of expression that do not exclude or denigrate anybody on the basis of gender, race, religion, culture age, disability or sexuality.

northpoint

a graphic indicator of the direction north on a map.

note identifier

a symbol or numeral used in text to refer a reader to a reference or to supplementary information in endnotes or footnotes.

note system

a system of referring to texts cited. It comprises a numeral in superscript within the text that refers the reader to full bibliographic details of the reference provided as a *footnote* (at the bottom of the page) or an *endnote* (at the end of the document). (Compare with *author–date system*.)

objectivity

unaffected by feelings, opinions or personal characteristics. Often contrasted with *subjectivity*.

online

connected to, under the control of, or accessible by a computer.

op. cit.

an abbreviation of the Latin phrase *opere citato*, meaning 'in the work cited'. It is used rarely in footnotes and endnotes.

orthophoto map

maps created from a mosaic of aerial photographs and overlain with information such as contours, transport routes and place names.

outline

to describe the main features, leaving out minor details. Alternatively, an outline can be a brief sketch or written plan.

paragraph

a cohesive, self-contained expression of an idea usually constituting part of a longer written document. Typically it comprises three parts: *topic sentence*, *supporting sentence(s)* and *clincher*.

paraphrase

to summarise someone else's words in your own.

parentheses

punctuation, written as round brackets, placed around a group of words (such as these) that is inserted into a sentence by way of explanation, but that is not grammatically necessary to the sentence.

peer review

a process of having peers or experts in the field review one's written work before its publication.

periodical

see *journal*.

pie chart

a circular-shaped graph in which proportions of some total sum (the whole 'pie') are depicted as 'slices'. The area of each slice is directly proportional to the size of the variable portrayed. Also known as circle graph.

plagiarism

presenting, without proper attribution, the product of someone else's work (for example, words, images or ideas) as your own.

population pyramid

a form of histogram showing the number or percentage of people in different age groups of a population.

poster

a large graphic display (for example, A3 size) comprising a piece of stiff card or some other rigid material upon which are affixed textual and graphic materials such as maps, tables and photos outlining the results of some piece of research.

PowerPoint

proprietary software to aid public text and graphics presentations. It includes word processing, outlining, drawing, graphing and presentation management tools.

precis

a brief summary of a piece of writing or a talk.

preface

the section at the start of a book or report in which the author states briefly how the book came to be written and its purpose. The preface will also usually include acknowledgments unless they are presented separately elsewhere.

primary data

information collected directly or obtained from first-hand experience (Compare with *secondary data*.)

prove

to demonstrate truth or falsity by use of evidence.

quotation (or quote)

a verbatim (that is, word for word) copy of someone else's words.

quotation marks

'inverted commas' that are placed around words reproduced exactly from someone else's speech or writing. They may be 'single' or "double" but their use must be consistent. Quotation marks may also be used to draw the reader's attention to a word that is somehow out of the ordinary. This device should not be overused.

range

the numerical difference between the upper and lower values in a range of data. For example, if the highest value in a series of property values is $950 000 and the lowest value is $400 000, the range is $550 000.

reference

source material used in the preparation of formal communication (for example, an essay, book review, poster or report).

reference management software

software for compiling and handling bibliographic information and for incorporating appropriate elements of that information into *reference lists* or *bibliographies*. A widely used form is *EndNote*.

references list

a complete list of works referred to or found useful in the preparation of a formal communication (for example, an essay, book review, poster or report). Usually, a list of references includes only those sources actually cited (that is, formally acknowledged). (Compare with *bibliography*. See also *author–date system* and *note system*.)

relate

to establish and show the connections between one phenomenon and another.

replication

repetition of a study to see if the same results will be achieved. A central tenet of the scientific method, the outcome may give credence to—or contradict—the results of the original study or experiment. (Compare with *corroboration*.)

representative fraction

a form of scale that expresses the relationship between distances on a map or diagram and distances in reality in the form of a fraction. For example, the representative fraction 1:10 000 (or 1/10 000) means that any one unit of distance on the map (for example, 1 millimetre, 1 inch or 1 metre) represents 10 000 of those same units in reality (for example, 10 000 millimetre, 10 000 inches or 10 000 metres).

review

to make a summary and examine the subject critically.

RF

see *representative fraction*.

scale

an indication provided on a map or diagram of the relationship between the size of some depicted phenomenon and its size in reality. A scale is used most commonly to provide a statement of the relationship between distances on the ground and distances shown on the map. Three forms of scale can be distinguished: a simple statement such as '1 centimetre represents 1 kilometre'; a graphic device that illustrates the relationship; or a *representative fraction* (for example, 1:15 000).

scattergram

a graph of point data plotted by their (*x,y*) coordinates.

scatter plot

see *scattergram*.

scholarly journals

journals that contain articles that have been written by, and for, academics or researchers in a field or subject area. They do not generally contain advertising material. Their articles often contain abstracts, research findings, data and extensive bibliographies.

search engine

software that searches a database (or databases), such as the *World Wide Web*, gathering and reporting on items that contain specified terms or characters. Common web search engines include Google, Yahoo! and bing.

secondary data

information collected by people or agencies and stored for purposes other than for the research project for which they are being used (for example, census data being used in an analysis of socio-economic status and water consumption). (Compare with *primary data*.)

semicolon

a punctuation mark (;) that indicates a longer break in a sentence than a comma, but is not as final as a full stop. It is now used mainly to mark the separations between long items in a list. It is not used to indicate the beginning of a list. (Contrast with *colon*.)

semi-log graph

a graph with a logarithmic scale on one axis and an arithmetic scale on the other. (Compare with *log-log graph*.)

sentence

a group of words that expresses a complete thought. It begins with a capital letter; ends with a full stop, exclamation mark or question mark; and contains a subject and a finite verb.

serial

see *journal*.

show

to demonstrate in a logical sequence why or how some phenomenon occurred or came to be.

sic

a Latin word that means, literally, 'thus'. It is used after a direct quotation if the quotation contains an error or a questionable statement. The word usually appears in italics and is placed in square brackets: [sic]. It is used to indicate 'this is the way it appeared'. If only one word is wrong in the quotation, [sic] appears directly after that word: 'He done [sic] great work on the field today.'

state

to express fully and clearly.

summarise

to present critical points in brief, clear form.

supporting sentence

that part of a paragraph in which discussion substantiating the paragraph's claim is presented. It provides the 'how' and 'why' examples to support the topic or prove the point of the paragraph (See also *topic sentence* and *clincher*.)

synthesise

to build up separate elements into some comprehensible whole. (Compare with *analyse*.)

table

a systematically arranged list of facts or numbers, usually set out in rows and columns.

third-person narrative

in the context of a research narrative (for example, a research report, journal publication or public talk), it involves constructing the narrative without reference to the researcher's thoughts, opinions or feelings. This consequently conveys a distanced and seemingly neutral, omniscient and 'objective' point of view. Third-person narratives have dominated in the presentation of 'scientific' research. (Compare with *first-person narrative*.)

topic sentence

that part of a paragraph in which the main idea is expressed. (See also *supporting sentence* and *clincher*.)

topographic map

a common, general-purpose map that typically depicts contours, physical (for example, rivers and peaks) and cultural features (for example, roads, churches and cemeteries).

trace

to describe the development of a phenomenon from some origin(s).

trend line

see *line of best fit*.

URL (Uniform Resource Locator)

a string of letters and numbers that make up the 'address' or location of documentary, image and other resources on the *World Wide Web*. For instance, the URL for the New Zealand Geographical Society is <www.nzgs.co.nz>.

viva voce

an oral examination used most commonly as a supplement to written exams, or to explore issues emerging from an Honours, Masters or PhD thesis.

web

see *World Wide Web*.

web page

a computer file that may be grouped with other related web pages to form a website. (Compare with *website*.)

website

a collection of related and linked *World Wide Web* files. See, for example, <www.iag.org.au>, which is the website for the Institute of Australian Geographers. Usually a website includes a beginning file, known as a *home page*, which sets out contents of the rest of the site.

Wikipedia

a web-based encyclopedia that is written collaboratively by volunteers and may be edited by users. Because it has no formal mechanisms for peer review, Wikipedia is often regarded as a questionable source of information.

word processing

the use of a computer to create, edit and print documents.

World Wide Web

Interlinked resources on the internet adhering to Hypertext Transfer Protocols for exchanging text, graphic, sound and other multimedia files on the web. Occasionally abbreviated as W3.

INDEX

Printed in Australia
11 Jan 2015
405016-A